THE LUCKIEST PLANET

Where Am I From? Why Am I Here?

MICHAEL CHAMBERS

© 2025 Michael Chambers All rights reserved.

Apart from any fair dealing for the purpose of study, research, criticism, or review, as permitted under the Copyright Act, no part may be reproduced by any process without written permission.

Unless otherwise stated, all Scriptures are taken from the New International Translation (Holy Bible. Copyright© 1996, 2004, 2007, 2013 by Tyndale House Foundation.)

A catalogue record for the book is available from the National Library of Australia.

ISBN: 978-1-7641963-1-4

Dedicated to my grandchildren: Jesse, Jacob, Blake, Ethan, Tyler, Josie, Riley, Lyla, Sophie, Jake and Abbie

About the author

Michael Chambers arrived by boat from New Zealand in 1969 and first worked as a jackaroo on the outback NSW station Mount Manara. His experiences over the next few years provided the background for the Tom Duncan adventure stories in his earlier books.

A fight with a Brahman bull in western Queensland—which he decisively lost—led to a new career with BHP merchandising companies in NSW. In 1989, he packed his wife, five kids, and a dog into a truck and van and headed for the Northern Territory, where he served as State Operations Manager.

A lifetime love of Aboriginal people culminated in a four-year relationship with the Yolgnu people of Arnhem Land, including taking a group of thirty Elcho Islanders on a memorable trip to Egypt and Israel. For fifteen years, he owned and ran the iconic NT fishing and camping shops, HAPPY MICKS TACKLE N TENTS—the name chosen by kids from the island. He was also a successful internet entrepreneur in the early years of the century.

In 2010, the family moved to Mildura on the picturesque Murray River. Michael and his wife Aileen both retired at age seventy—she from over twenty-five years in aged and palliative care, he from a long stint in retail. Together they've raised five children and now enjoy eleven grandchildren.

Drawing from decades of diverse experiences across Australia's landscapes and cultures, Michael writes most days, bringing a unique perspective to life's biggest questions.

Contents

an Introduction To The Luckiest Planet.................................2

Prologue 9

   Chapter 1: What is the Bible? 15

   Chapter 2: A Story of Science 24

   Chapter 3: Beginnings..43

   Chapter 4: Seven Days that Divide the World 49

   Chapter 5: Genesis Day One 56

   Chapter 6: Genesis Day 2 64

   Chapter 7: Genesis Day 3 69

   Chapter 8: Genesis Day 4 74

   Chapter 9: Genesis Day 5 82

   Chapter 10: Genesis Day 6 85

   Chapter 11: What about Darwin? 92

   Chapter 12: Mico Evolution and Nanotechnology ...98

   Chapter 13: Devolution and Secrets of the Cell 102

   Chapter 14: Where Did Adam Come From? 112

   Chapter 15.: What About the Dinosaurs? 117

Part 2: The Story So Far 124

   Chapter 16: The Story So Far

   (... it's all about religion, stupid!) 126

   Chapter 17: The Horror of War 135

   Chapter 18: The Stupidity of Racism! 151

Part 3: Why Am I Here? 169

   Chapter 19: The Pale Blue Dot. 170

Chapter 20: Here's a Real Climate Catastrophe! 179
Chapter 21: For the times, they are a ' changin'! 188
Chapter 22: It's All About Energy! 201
Chapter 23: The Abundant Planet 222
Chapter 24: The Resilient Planet 230
Chapter 25: The Golden Calf 236
Chapter 26: The Third Day 243
Chapter 27: In Conclusion. 246

From a distance, the world looks blue and green
And the snow-capped mountains white
From a distance, the ocean meets the stream
And the eagle takes to flight
From a distance, there is harmony
And it echoes through the land
It's the voice of hope
It's the voice of peace
It's the voice of every man
From a distance, we all have enough
And no one is in need
And there are no guns, no bombs, and no disease
No hungry mouths to feed
From a distance, we are instruments
Marching in a common band
Playing songs of hope
Playing songs of peace
They're the songs of every man
Julie R. Gold

An Introduction To The Luckiest Planet.

Recently, after a family wedding, it was suggested by my many grown children and numerous grandkids that I write them a book about what I know and have learned during my time on this earth. A worldview that they could understand and relate to.

Had this happened decades ago, it would have been quite ridiculous for an older man approaching eighty to entertain such a project. But during those years, life has changed so much that I find it hard to figure out where to begin.

The idea of writing a new book was not so silly. Indeed, since I retired at age seventy, I have published two historical novels about Aboriginal Australia and European settlement. WALKING AMONG THE STARS is set in the NSW outback near Lake Mungo, where I met my wife, whose family farmed there for over 150 years. I tell the story of the Ngiyampaa people, who have lived there for thousands of years. Then, drawing on my experiences of living with the Yolngu people of Arnhemland, I recently published MAKARRATA, THE AUSTRALIANS OF ARNHEMLAND. So, writing comes naturally, but their idea for this book was different.

Although I left school at sixteen and had no formal education beyond that, I love learning and have an inquiring mind. For instance, I read my mother's encyclopedias from cover to cover before I was twelve. And I have been learning since, not at university but in the school of life and experience. Knowing of my love for natural science and the Bible, they asked me to tell how those two subjects complement each other, not in competition as many things. So here are some stories, based on extensive research, about THE LUCKIEST PLANET. WHERE AM I FROM? WHY AM I HERE?

I was introduced to this wonderful planet we call home when I was young. I was born in England shortly after the war, one of the first "baby boomers," in a small, isolated village tucked away on the edge of the remote Yorkshire moors. My parents had both been in the war; Dad was a Spitfire mechanic in the Air Force, and Mum was a WREN, Women's Royal English Navy, a personal secretary to a great Admiral, no less! As many did in those times, we lived with my grandparents, loveable Christians of the Quaker faith. I was raised in the Quaker faith until I was in my teens. Then, like many, I rebelled and became a devout atheist!

When I was five, my father said goodbye and took a passenger ship to New Zealand on the other side of the world. There, he built a house for our family to live in. Several months later, my mother, sister, and I took a train to Southampton and boarded the SS RANGITIKI, an immigrant ship. They called us "ten–pound poms," for that was the boat fare to our new home. So began my first voyage of discovery of this vast, glorious place we call home.

I soon got a taste of just how big! The voyage took six weeks – across the Atlantic to the West Indies, as Christopher Columbus did. Through the Panama Canal that divides North America from the Southern continent. Over the vast Pacific Ocean to Fiji and Pitcairn Island in the wake of the great James Cook, the eighteenth-century Yorkshire navigator and explorer. As a five-year-old boy, I, too, was on a grand voyage of discovery, weeks and weeks at sea with only the sun, moon and stars for companions.

It would be nice to say that we lived happily ever after, but that did not happen. After a few short years in New Zealand, my parents separated and divorced, a rare occurrence in those days. My mother bundled my younger sister and I onto a massive BOAC Constellation plane. We flew back halfway around the world, across Asia, India and Europe, to my grandparents' home, now in County Durham near the Scottish Border. By the age of ten, I had circumnavigated the world. I acquired three accents: a Yorkshire brogue, a north country Geordie, and an Antipodean twang from New Zealand. My new friends thought I talked funny but accepted me for my novelty status as a great world traveller!

But wait! The vast oceans were not done with me yet. Eighteen months later, we took another train to Southampton and boarded the SS DOMINION MONARCH. This time, we took a different route; across the Bay of Biscay, down the coast of Africa to Cape Town, and then across the Indian Ocean to Australia and New Zealand. As a ten-year-old, I remember this experience vividly and some life-changing events that had a huge impact. There was a small group of boys my age. We played and talked constantly, free to roam the enormous vessel at will, constrained only at mealtimes, when we would be herded into the lower deck restaurant to sit with our families and behave. Especially on those rare occasions when we were invited to sit at the captain's table. He was a grand old man of the sea, bewhiskered and growly, and would often deliberately seek out our gang and regale us with tales of life on the seven seas. So, a lesson; old men's stories *can* be informative and entertaining!

One sultry tropical afternoon, not long after the Crossing the Equator festivities, we played with our new toy racing cars our parents had liberated from the ship's shop. We were on the shady rear lower deck, a calm, pleasant place to spend an afternoon. The ship's white-water wake stretched all the way to the horizon. Suddenly, my mate squealed in pain as the deck became cluttered with numerous colossal fish. He had been hit in the back by a flying fish. This unique creature leaves the water at speed, gliding for considerable distances. They were familiar, but we had never seen them this close. We soon discovered that by crawling on our hands and knees, we could shove these silver, slimy, struggling creatures back over the side and into the white wake water below, where they again took off, flying away at prodigious speeds. An excellent introduction to the animal kingdom and the mysterious mechanics of flight. And a treasured memory.

We berthed at Cape Town on the southernmost tip of Africa. My mother, sister, brother and I wandered the streets, entranced by the tiny squirrels that scampered and played in the numerous parks. Walking down the main street, wandering and window shopping, a family approached us on the pavement. A mother and two daughters, all dressed in pink gingham-checked frocks, a gorgeous family, but the blackest people we had ever seen! As they approached, wide-eyed and with gloriously white shining teeth, they jumped off the footpath onto the street and bowed as we went past, still happy and smiling. As Africa slid gently back over the horizon that evening, we sat on the promenade in deck chairs, sipping beef tea served by waiters in shiny white jackets. I asked Mum what had happened with those black people. She told me about the horrors of apartheid. Of racism and the hatred of people different from others. Of the Holocaust and the genocide of the Jewish people by the Nazis. Whom both my parents had fought against! Another important memory.

Many years later, in 1963, I was playing cricket at school on a sunny summer afternoon. We were a multi-racial mob, with Europeans mixed with Maoris and many in-betweens. Someone had a transistor radio, an essential item in those days now long consigned to the scrap heap of progress. As we gathered around, we heard a deep American voice, strident and urgent. His name was Martin Luther King Jr, a black Negro minister and leader of the civil rights movement.

> "I say to you today, my friends, so even though we face the difficulties of today and tomorrow, I still have a dream. It is a dream deeply rooted in the American dream.... I have a dream that my four little children will one day live in a nation where they will not be judged by the colour of their skin but by the content of their character."

Martin Luther King Jr 28th August 1963

I was of a generation, one of the first teenagers. We had lived our young lives in the shadow of the Cold War, the Cuban missile crisis, and the constant threat of nuclear war. But as a generation, we rose to Doctor King's challenge. This was it! Righteousness had prevailed, and we could all dream of a fantastic future.

Unfortunately, there are many today who do not share that dream, many who *should* but *actually* oppose it. So, we will talk about that, too.

As I have said, I have studied both science and the Bible. I have listened to scientists and theologians who, like me, see no conflict between the two. And I would like to share some of those thoughts and ideas with you.

However, understanding the world requires more than scientists and theologians. It needs philosophers, poets, painters, and people who dream and envision. Even politicians of that mode. One such is the late great British philosopher Sir Roger Scruton, who wrote:

> "Anybody who goes through life with an open mind and heart will encounter moments that are saturated with meaning but whose meaning cannot be put into words. These moments are precious to us. When they occur, it is as though, on the winding ill-lit stairway of our life, we suddenly come across a window through which we catch sight of another and brighter world, to which we belong but cannot enter. There are many who dismiss this world as unscientific fiction. I am not alone in thinking it real and important."

So, I invite you to share this next journey of discovery with me. Together, we can explore ideas and thoughts, dreams and visions. We may not always agree, but that's fine – we do not have to. Isn't that the point of discovery, of setting off on a voyage of exploration together?

I remember sitting on the top deck of the great passenger ship, full of migrants excited by a new life in a new land. We had seen sunrises and sunsets that were incomparable in their beauty. And now, we gazed into the glorious star-filled sky, as far from city lights as you can get on this planet. It was so dark yet so bright. My mother was a well-educated modern woman, a woman of her time. She taught me a lot. As we gazed at the outrageous beauty above, she spoke of the moon, the planets and the stars. Of how they were suns, like our suns. Of how there were hundreds of thousands of them. Of how the Universe and all creation were static, a place of wonder but with no beginning and end. This has been a common view throughout the ages and is still held by some scientists today. And she spoke of the little things, the smallest of things like the atom. As I said, she was a well-educated woman of her time.

But nearly seventy years on, how have things changed? We now realise that yes, there was a beginning, as the Bible records, and therefore there will be an end. And yes, there are some stars out there. Still, most of those shiny lights are actually enormous galaxies, far grander than our own Milky Way, and there are billions and billions of them. And the small things. The cells that form our bodies; trillions of them. They are controlled internally and activated by various minute complex biological machinery. It's called nanotechnology, and the recent discoveries in Quantum Science have changed our lives. And it's amazingly complex and wonderful. Did you know that?

So, let's see if we can "*.... catch sight of another and brighter world, to which we belong but cannot enter.*"

Because maybe we can catch more than just a glimpse. Perhaps we can enter it. Now, there is a worthy voyage of exploration. So, join me, and we will enjoy the journey together. Maybe it is to learn why your life and mine make this planet we call home the luckiest in the Universe. And maybe, just maybe, we will glimpse that other brighter world to which we all belong.

Prologue

A story from Mount Sinai: about 3500 years ago
From Exodus 38:21.

> These are the records of the tabernacle of the testimony, as they were recorded at the commandment of Moses, the responsibility of the Levites under the direction of Ithamar, the son of Aaron the priest.

The fire that had blazed brightly all night was fading to a glow. Ithamar pulled his heavy woollen cloak around him in response to the early morning chill, which had woken him from a light sleep, head slumped exhausted over his knees. In the east, there was a faint lightening of the sky as the once-dominant stars began to fade. Not long now, he thought. He stood with his staff and spear, a lithe, active young man about to enter the prime of manhood. He could vaguely see the outline of sticks and thorns of the enclosure and sensed rather than saw his flock of sheep and goats still soundlessly at rest.

His father, Aaron, had negotiated grazing time at this remote well with Jethro, priest of Midian. Today, he would begin returning to the mountain where the tribes of Israel were camped.

He stretched generously and wandered over to the well, scooping cold, precious water from the trough into his face. In the faint first light of day, he was a tall, well-built figure with a dark, manly face, intelligent, deep brown eyes and a shock of unruly black hair that would attract his mother's attention when he finally arrived home.

Ithamar was the youngest of Aaron's four sons, the other three destined for the newly created Priesthood, of which his father was soon to be Head Priest. He was also the elder brother of Moses, and in Egypt, he was his mouthpiece and orator. For as long as he could remember, his father and uncle had gathered around the smoky tallow lamps late into the night, quietly plotting the exodus of the Hebrews from Egypt. As the youngest, he was in charge of the family's flock. This was how it had always been, but his father had taken him aside as he left the Hebrew camp of some two million souls.

"This will be your last trip, son," he explained. "We have other work for you. May Adonai be with you. Go in peace."

That had been two weeks ago, and he had indeed had a good trip. No wolves or lions this time, though he had killed both in the past. The Midianites had done an excellent job keeping them out of their grazing grounds. All the shepherds of Israel were masters of the slingshot, and they were all proficient with a deadly spear, taller than a man. Unlike the main camp, he could forage for ample food and grazing for his flock. Unlike those back at camp; Lord Adonai fed them twice daily with massive schools of quail that could be roasted during the day. And every morning, a manna fell like a frost on the ground, easy to gather and nutritious. But there was none of this miraculous food for Ithamar.

He became aware that his flock was awake and exhibiting signs of nervousness, bleating, and anxiousness. And then the unmistakable roar, the challenge of a bear on the hunt! He squinted into the gloom, searching where the challenge had come from. Finally, he made out the faint but unmistakable figure of a great brown bear, one of his most dangerous enemies.

He ran back to the fire, his mind racing. Would the giant animal respond to a challenge, or would it tear down the stockade and begin to slaughter his flock? He grabbed his spear as he reached the dying fire; it was still too dark for the sling. And he let out a roar of his own, one intended to challenge.

"Ooyee! Ooyee!!"

He saw the animal pause, head up, testing the air. It stood on its hind legs, taller now than he. He sensed they could finally see each other. It began ambling unhurriedly around the enclosure towards the shepherd. That was the desired outcome, thought Ithamar.

Swiftly, he raked the remaining glowing coals, sweeping them in a large arc, embers now covered in desert sand, no longer glowing. And then, he just stood, patient. Confident.

It could be said that the Hebrews were a brave lot. Even stubborn. Had they not entered the narrow channel between two great banks of water? As a nation, they defied the pursuing Egyptians, who intended to use their chariots to bring genocide to the sons and daughters of Israel. And then, when all were safely across, with a mighty hand, the Lord, Adonai, had caused the waters to close over the now terrified and doomed pursuers. Every man, woman, and child who came through those waters had bravery woven into their character. A brave, stubborn, stiff-necked people. Some would say stupid and ungrateful even!

But not Ithamar. He had spent his formative years here, in the desert with his flock. Never lonely, he had developed a close relationship with his God, constantly singing his praises. Building his faith. A faith undeterred now by fear. He did not fear this remarkable creature, but he did respect it. He anticipated what it would do, how it would behave. And so, it came to be.

The bear approached the young man slowly, arrogantly, on all fours. Stopping occasionally to rear up on his hind legs and repeat his challenge. But always closer and more dangerous.

Ithamar stood his ground, spear by his side. He rebuked the great animal directly, loudly, explaining in detail what would happen. They closed to within a few feet, the bear roaring constantly, rearing again. Dominating the lone shepherd. Foul breath now apparent. He spoke a final prayer. The bear lunged forward, threatening death. As the dormant fiery coals burned through both front pads, it released a primal scream and reared up in fright. Ithamar struck instantly. The spear pierced the heart, and the creature died instantly.

***

Late in the evening of the following day, he approached the camp at the head of his flock, who were trained to follow wherever he went. To his surprise, his father Aaron was there to meet him, perched on a large rock just outside the camp. Two of his younger cousins were with him. They embraced, and his father asked about the trip.

"There was a great bear," was all he said. "What's going on?" he indicated to the two boys.

"Moses wants to see you. Immediately!" he added.

'Can I wash and change?" he asked.

"No. The boys will deal with the flock. Just point out the firstborn for sacrifice. We will do that in the morning. Now come with me."

As they walked through the vast encampment, Aaron explained to his son what had been happening. While Ithamar was away, Moses finally came down from the mountain of smoke and fire, carrying the Tablets of the Law as Adonia gave. But in the meantime, some of the leaders had revolted. They collected a considerable amount of gold from the people, who were also grumbling about their situation. They made a Golden Calf, an idol of one of the gods of Egypt, from where they had just escaped. Moses was horrified. He broke the sacred tablets on a rock and ordered the leaders to be killed.

"Is that why there is so much mourning and weeping in the camp?" asked Ithamar.

"Yes," replied the older man. "But then Moses had three thousand men involved executed as well. It wasn't very pleasant, but it was just. That's why many are grieving."

Despite the attempted rebellion, there was activity everywhere. Artisans were hammering and shaping gold and bronze. Weavers created great swathes of cloth: gold, black, purple, and scarlet. Craftsmen were creating poles and planks from seasoned acacia wood. The camp was a hive of activity.

"What's all this about, father?" the young man asked.

Adonai has instructed us to make Him a Tabernacle, His dwelling place on our journey. Moses will explain.

They approached Moses' tent at the foot of the mountain, near the freshwater spring Moses had summoned from a rock with his staff. They passed the guards and entered a comfortable living space, big enough for meetings, dark and murky, thick with the smoke of many lamps.

Moses sat on a pile of cushions at the rear of the tent. He motioned them to sit opposite. It was a few weeks since Ithamar had seen his uncle. Did he seem older? No, it was not that. It was his face. It was bright in the gloom of the tent. It was almost ... glowing.

Ithamar apologised to the older man for not being clean and appropriately dressed. Moses waved him aside gruffly.

"Tomorrow, I return to the mountain to inform Adonai of the tragedy that has befallen us. But, in the meantime, I have a job for you," he said, indicating the pile of papyrus, pens, and ink on the floor beside him.

"Surely, uncle. Whatever you decide."

"I want you, Ithamar, son of Aaron, to be my scribe. To account for the materials of the Tabernacle, the gold and precious stones, woven cloth, and all that was given willingly by the people. A record for the people, open and transparent. As well you will write for me. I have much to tell and little time to tell it." He paused and looked intently at the young man before him. "Do you accept?"

"Yes indeed, Moses. Where do we start," he asked, taking up writing materials, pen poised,

Moses looked fondly at the filthy, unkempt young man before him. "Remember. Every jot and tittle! Start here," he said, taking a deep breath. "In the beginning, God created the heavens and the earth."

[Learn more here.](#)

CHAPTER 1

What is the Bible?

Perhaps it is better to start with what it is not.

It is not a rule book. It is not a book about laws. It is a series of 66 books, the most extraordinary storybook ever, written by real people, about actual events. It was written by different authors, all Jewish, over a period of about fifteen hundred years. That's a long time. During that time, the Hebrew culture and language changed. Therefore, it is excellent for teaching about God, His people, and life. Even a few thousand years later.

The first five books are called the Torah, and Jewish tradition says it was written by Moses. Notwithstanding that, the origins of the Bible go back long before even writing was invented, back to the times when history and culture were passed on orally, by mouth. It is true; these stories were a myth. But, unlike the fictional Star Wars: Luke Skywalker was a myth – until he turned out to be true and became a legend.

I was involved with the aboriginal Yolngu people of Arnhemland in Northern Australia some years ago. For thousands of years, they have retained their language, land, and much of their culture. However, only in the last century did they come into contact with the Christian missionaries and Western civilisation. And writing. For all of time, their culture and worldview had been recorded orally and in art form. By memory, not by writing. In my recent book _Makarrata_, I use the papers and writings of an anthropologist who spent seven years with these people, learning their language and culture, listening to their stories and recording them for prosperity. And now, a hundred years later, they pass on their culture orally *and* in written form. The story of the Yolngu people of Arnhemland.

The Bible is the story of the Hebrew people of Israel, known as the Jews. It, too, started as an oral history before writing was even invented. We call that "myth," and some modern scholars suggest it is fiction. However, as we learn from the Yolngu people, myth *can also be* oral history. It can carry *the exact authenticity* of written history.

About 3500 years ago, oral history began to be transcribed carefully into the new Hebrew language. [Modern archaeology suggests that this was the first written language](#) using a unique alphabet invented while the Hebrews were still in captivity in Egypt.

That was long ago, back at the dawn of recorded history. Even so, sufficient elements and discoveries from modern-day science, such as anthropology and archaeology, suggest authenticity to the work.

But it depends on your worldview. Your viewpoint. What do I mean by that?

Imagine you are a satellite, not hovering in one place over the Earth as some do. No, this satellite sits on the far side of the moon, the side that is never revealed to Earth. What is your view of the moon? You think the moon is a dark, forbidding place with no geological features. And from the point of view of that satellite, that is understandable. But what if the satellite suddenly hovers over the bright side of the moon, facing Earth? Aah! That is different. Now the moon is a place of light, with mountains, valleys, craters and other interesting geological features.

You see, your opinions are influenced by your viewpoint. In this case, light side or dark side. In this book, we will spend much time looking at *duality*. That is a big word, but it means looking at things from two perspectives. Or it can mean something having two completely different characteristics or functions. Actually, our lives are full of duality. And as we shall discover, so is science. And so is the Bible.

The Bible is the story of Israel, of the Jewish people. Though often not acknowledged, it is an entirely Jewish book by Jewish authors.

On the other hand, it is also the story of God and His relationship with Creation. With all people created In His Own Image!

That is duality.

So, depending on *your* viewpoint, it is either the inspired word of God, as faithfully recorded down through the ages. On the other hand, you may think it is not inspired by God and is, therefore, mainly a myth. I am writing this book from the former viewpoint that the Bible is the inspired word of God and is backed up by sufficient evidence, from Creation to the Cross, to legitimise my faith in that statement!

The Bible is not only the most-read book in the history of the world. It is also the most written about and the most researched. Recently, numerous discoveries in the Middle East and beyond authenticate and support this ancient book and its claims. It is beyond our scope to explore those amazing findings. So, I have chosen one story, one example of a discovery that is breathtaking in its revelation of the ancient world and proves beyond any doubt the accuracy and age of the Bible. It deals with Ithamar and the sons of Aaron.

The following story is taken from the documentary The Mystery of the Silver Scrolls.

In the modern city of Jerusalem, the capital of Israel, there was a quiet, secluded hillside that would have been outside the walls of the ancient city of King David. Ketef Hinnom is separated from the city by the road to Bethlehem. Since the earliest times, travellers of all descriptions have passed through here. It is, in fact, the site of an ancient burial ground dating back to the time of King Solomon's first Temple. Since it was ravaged by grave robbers, it had never been adequately excavated or studied. In 1979, Doctor Gabriel Barkay was widely regarded as the Dean of Biblical Archaeology.

Here is how he tells the story in his own words, in the documentary: The Mystery of the Silver Scrolls

> I was born in the Jewish ghetto of Budapest, Hungary. Actually, nobody should be born there. In January of 1945, the Russian Army entered the ghetto of Budapest, and since then, I've been a free man, and I'm alive. I guess that between myself and death in the Nazi concentration camps, there is about a week or so. It was the end of the war. The Germans were in total disorder already, and they forgot about us. I owe my life to the Russian Red Army and to Joseph Stalin. Many owe him their death. I owe him my life.

It was a miraculous salvation. Gabriel's family moved to Israel, and he grew up in Jerusalem with a keen interest in Biblical archaeology, particularly ancient Jerusalem. He left Tel Aviv University wanting to earn his Doctorate in a site he had carefully chosen – Ketef Hinnom. Not knowing that his discoveries would change and excite the world!

As was common then, he had little financial support for the project. He started the dig with a handful of young boys from a nearby school.

After instructing them about the delicacy of items they may find, they carefully excavated what turned out to be an extensive graveyard. After only three days, they discovered a beautiful burial cave belonging to a wealthy family. Gabriel realised it dated to the First Temple period, about 2800 years ago. He was disappointed to find no items of value or interest, so thorough had been the looting. However, it was a typical example of a wealthy family's burial practices. Inside the small cave were slabs of rock where bodies could be laid for the first part of the ancient ceremony. Headrests were carved out of the rock where each body could be laid. Underneath was another chamber where the bones would eventually be laid to rest. It was a significant find, and Gabriel decided it should be photographed for publication.

One of his young helpers was a boy called Nathan, an unruly child, a nuisance, challenging to work with on a site as serious as this.

"Nathan," he said, showing the lad the grotto. I want you to take brushes and clean this whole place out. Tomorrow we will take photos. So, it must be spotless and cleaner than your mother's kitchen. Understand?"

The boy nodded, and Gabriel left him, happy to have found the rascal sufficient work to keep him occupied.

Sometime later, while Dr Barkay worked with his team on a lower level, he felt a tug on his shirt from behind. He turned to find Nathan standing behind him with a beautiful alabaster jug in his arms, grinning wildly. Gabriel was horrified and gently stored the precious find before investigating its source.

Nathan had finished his work and was bored. Being who he was, he found a hammer and began hammering on the chamber floor. Before long, it cracked and opened up, revealing another chamber below.

Gabriel eased himself down into what was an ancient burial chamber. By the light of his lamp, he gazed at a room that had been designed never again to be seen by human eyes. There were skulls and skeletons everywhere, 95 to be exact. Treasure, burial gifts, jewellery, trinkets, gold, silver, and precious stones. Pottery and vessels of alabaster and ivory. Astounding treasure, untouched and unseen for thousands of years. Gabriel was both astonished and excited. This was unprecedented.

He summarily dismissed his young crew of workers with thanks. He had the University send out a team of graduates who would take care of and value any findings they made. They were sworn to secrecy; the possibility of looting loomed until they could get security. Gabriel gradually realised the significance of the find, possibly the find of the century in Jerusalem.

But the best was yet to come. The next day, they cleared a ledge in the chamber with tiny brushes carefully, gently, and slowly. Suddenly, a yell. A small cylinder, the size and shape of a cigarette butt, about 2 centimetres long and half as much in diameter, with the look of aged silver. Then another one, about half the size. Gabriel quickly identified them as amulets. He pointed to a small loop on each, where they could be hung around the neck. Inside would be a scroll, also made from silver foil.

It was a significant find, probably around 2800 years old. But the two objects would not give up their secrets easily. Would they risk destroying the precious items by trying to extract the scrolls? It was too enticing. Gabriel decided they would take the risk.

The two silver scrolls were sent to a laboratory in Leeds, England. After a time, they sent them back untouched. They were not willing to take the risk. So, Gabriel sent them to a world-renowned laboratory in Berlin. They came back – the same message. Gabriel tried himself, using heat and flame, but to no avail. Finally, after three years had passed, Professor Joseph Dodshinov from the Israel Museum offered to help. Painstakingly using a combination of acid and acrylic emulsions, he presented Doctor Gabriel Barkay with two small silver parchments. And he was thrilled to see they both had writing on them. He could make out the Hebrew character YWAH, the sacred Hebrew name of God. This was exciting. But no one was able to translate the rest of the letters. And it seemed that was the end of the story. But no, not quite!

Many years later, in 1988, the two silver scrolls were sent to the Israel Museum for a unique display. Ada Yardini, a scientist there, noticed that not only was the word YWAH (the Hebrew word for God) there, but it appeared three times. She phoned a friend, a noted Bible scholar, who immediately identified the verse. "That's the Priestly Blessing," he said. "From the book of Numbers. That's the only place in the Bible where YWAH is used in three consecutive verses." And so it would be translated.

Letter by letter, word for word, it faithfully recreates the blessing given to Moses by YWAH to bless Aaron, Ithamar and his three brothers, some 700 years before. Three priests and Ithamar, the scribe!

The Luckiest Planet

The Lord bless you
and keep you;
the Lord make his face shine on you
and be gracious to you;
the Lord turn his face toward you
and give you peace
Numbers 6:24-26

We will let Doctor Gabriel Barkay explain in his own words, from the Documentary: The Mystery of the Silver Scrolls

"That blessing is, first of all, beautiful. The language is so beautiful, but the Hebrew is lost in most of the translations. It starts with the Blessing of God and ends with the word "Shalom" meaning perfection, meaning harmony between all components of Creation. Today, for us, Shalom is peace, but it is much more than that. In any case, the significance of that word is the ultimate peak of the aspirations of every believer. This prayer is still in use in Jewish and Christian prayer, and on the high holidays, it is read to an enormous public next to the wall of the Temple here in Jerusalem. It is used in public ceremonies such as the inauguration of American presidents. The Priestly Benediction is read by Jewish fathers to their sons and daughters on the eve of the Sabbath, as they come from welcoming the Sabbath. These are the first Hebrew words I remember from my early childhood. These are the words with which my late father blessed me. This is a kind of closing of a circle for me. The text is a beautiful text and it is also of personal importance to me."

What a remarkable story from a man born into the horror of the Nazi Holocaust.

The Silver Scrolls contain the oldest biblical inscription ever found. When they were made, the prophet Jeremiah was still alive. Solomon's Temple was still standing, and the heirs of King David were still on the throne. They are 400 years older than the famous Dead Sea Scrolls.

Summary:

> *The Bible contains amazing literature, beautiful poetry, and memorable stories.*
>
> *It is also historically factual, reliable and verifiable.*

CHAPTER 2

A Story of Science

Once upon a time. A long, long time ago. Many generations *before* Moses led the Hebrews out of Egypt.

The day was gently coming to a close on the alluvial plain between the two great rivers to the North and East of Sinai. The sun was long gone, and his family was asleep inside a cluster of huts. Crude shelters of branches and thatch that kept out the worst of the wind, rain, and dew. Perched as they were on top of a slight rise, they were high and dry above all but the worst floods, welcome and frequent as they were.

The old man was the patriarch of a tribe of twenty-seven souls: two younger brothers and their families; his two wives, their children, and, most recently, a couple of grandchildren. He grinned inwardly. It was a fine family to comfort and nourish a man in his old age. And he was older than most.

He sat comfortably on a huge log others had carried here for this purpose. That was some years back and is now a source of considerable comfort. It was far enough from the roaring fire not to burn but close enough to ease the nagging pain in his joints. It was a good log.

He had led his family to this place while he was still a young man. They were among the first to settle and become farmers, growing crops from seed in the rich, fertile soil of the floodplain. As a young boy, he had been taught to hunt by his father. Now, he and his family were farmers bound forever to their crops and reliant on the rains for their food. He still hunted in the old style, but his sons would be more connected to their crops than to their search for game. Indeed, they now kept animals themselves to graze and grow fat or pull the plough and labour in the fields.

The old man spent most of the day with a grandson by the river. The women had made him a new net, and he demonstrated to the lad how and where to cast skilfully for the plumpest, tastiest fish.

But this was his favourite time of day. It was quiet, and he could think, which was important. He studied the heavens above, sharp and bright in the clear air. He searched for *his* stars, those his father had pointed out by name and his father before that. He knew them all; where they were, where they were heading, when they would rise and set. He was particularly interested in those that moved gracefully across the sky, a fire trail in their tail. He knew they were a different kind of star and named them accordingly. Like most men of his time, he appropriated differing religious values so he told stories about each of the familiar heavenly bodies. And they numbered in the hundreds!

He was a man of considerable curiosity for the world around him. The slow, elegant, predictable flow of the seasons. The majestic beauty of a summer storm, wandering across the vast plains, at once leaving both refreshing rains and dangerous fires as the lightning bolts cast their angry fire onto the defenceless Earth.

Whenever he butchered an animal, fowl or fish, he noticed the entrails, ascribing with wonder names and purpose to each organ: the heart, lungs, kidneys, liver, and brain.

In this time and age, he was not the only man to be thus engaged; now, many were like him. Thinkers, dreamers, philosophers, curious students of the natural order and the world they lived in. He sensed something beyond that which he could see and touch. Particularly when gazing deep into the night sky, there was a stirring deep inside that he could neither recognise nor identify. It was an uncomfortable restlessness, a realisation of something else, something more profound. Greater even?

Speaking of these things Ecclesiastes 3:11 says

He has made everything beautiful in its time. He has also set eternity in the human heart, yet no one can fathom what God has done from beginning to end.

Quite remarkably, this nameless old man and those like him were the world's first "natural philosophers." Or scientists, as we call them nowadays.

❋ ❋ ❋

What is Science?

According to Stephen C Meyer in The Return of the God Hypothesis,

> "Science is the pursuit and application of knowledge and understanding of the natural and social world following a systematic methodology based on evidence."

In other words, Science is the study of Nature, the natural world, and everything in it. It is a system, a universally accepted method of arriving at a theory supported by empirical evidence, that can be both tested experimentally and replicated. (Empirical means evidence based on experience or observation.)

The scientific arena has grown so much in recent years that it extends to studying the universe from as far back as the Big Bang some 14 billion years ago to the engineering of the smallest nano-particles so minute and difficult to comprehend with the human mind.

> Contrary to popular belief, there is no deep divide between the Bible and Science. One is a system of study, and the other is a history of the Jewish people, resulting in the rise of Christianity. Thetwo are not mutually exclusive. As we shall see, both were factors in the growth and success of the discipline we call Science today. Stephen C Meyer says
>
> "…… the claim that science and religion are completely separate often conceals the assumption that science deals with reality and religion with things like Santa Claus, the Tooth Fairy, and God. The impression that science deals with truth, whereas religion deals with fantasy, is widespread. No one convinced of the truth, inspiration, and authority of Scripture could agree with that."

So, how did we arrive at this incredibly compelling field of human understanding and endeavour as a civilisation? How did mankind go from an old man sitting by a fire gazing wistfully at the moon – to another man walking *on* the moon? In just a few thousand years, the blink of a cosmological eye?

It's a good story.

✳ ✳ ✳

Where did Science begin?

It is widely accepted that the modern scientific method grew organically from advances in Western European society in the 16th and 17th centuries. But why there? And why then?

In his book, The Return of the God Hypothesis, Professor Meyer, scientist, author, and commentator, describes it like this:

> "…… he observed that the material necessities for conducting Science existed in many well-developed cultures. The Egyptians erected great pyramids, palaces, and funerary monuments. The Chinese invented the compass, block printing, and gunpowder. The Romans built great roads and aqueducts. And the Greeks had great philosophers, some of whom studied Nature extensively. Yet none of these cultures developed the systematic methods for investigating Nature that arose in Western Europe between about 1500 and 1750
>
> Let's think about what is needed for the viable birth of Science. We see first that it needs a reasonably well-developed society, so that some of its members can spend most of their time just thinking about the world, without the constant preoccupation of finding the next meal. It needs some simple technology so that the apparatus required for experiments can be constructed. There must also be a system of writing so that the results can be recorded and shared with other scientists, and a mathematical notation for the numerical results of measurements. These may be called the material necessities of Science."

So, we can deduce that those material necessities for the emergence and growth of Science were not only available but were unique to The West. How so? The answer may surprise you!

In the fifth century, the once mighty Roman Empire was already under pressure from the burgeoning nations of the East. And it was being attacked by Germanic barbarian tribes from the West. The Romans had already lost Britain and all of Europe. This collection of tribes now stood at the gates of Rome itself. They were brutal, war-loving people, loosely governed by tribal warlords and constantly changing loyalties and treaties. Yet, when facing the common enemy, they fought as one. Eventually, Rome was overrun due to centuries of insidious Barbarian growth and pressure. But in its final anguish, the Romans bequeathed a great and precious gift to their conquerors that would change not only the heathen tribes but the whole world. When they finally returned to their native lands, they carried in their hearts the new religion; Christianity.

Something about the new faith appealed to the Barbarians' deep love of freedom, both individual and tribal groups. Instead of the mythical gods of fire and thunder, fashioned in the image of the sun and moon, animals and sea creatures, here was a man, like themselves, a God but also kind, gentle, and above all, a lover of freedom. The great Christ, King of Kings and Lord of Lords, was also a commoner, the son of a carpenter, who knew violence and death but rose again despite it.

Seek not in courts or palaces,
Nor royal curtains draw
But search the stable
See your God
Extended on the straw
The Shepherds Prayer, William Billings

So, these barbarians learned to love their neighbour, not just their tribe. Christianity appealed to their ingrained sense of equality and freedom, of the rights of the common man. It was a rough and ready religion in their hands, but over time, it changed the character of Europe and most of the known world. Centuries later, this sense of freedom would lead to a small, lonely field outside London, England. There, a king would kneel before a group of rebel lords and barons and submit himself, his heirs and successors to a list of outrageous demands. They all signed the Magna Carta document, which stated that monarchs would henceforth be subject to the law and not above it. Common law. This, too, had a significant impact on civilisation.

All across Europe, great abbeys and cathedrals appeared, and many were the works of architectural geniuses. Towns and cities emerged. Artists, authors and philosophers flourished alongside those studying the natural world. With the general softening of the culture, society changed – for the better. Those like the Saxon Queen Matilda made considerable advances in reducing slavery. Conversion to the new religion was generally made by personal choice. Women were better treated as equals. It became common for a woman to marry for love, by choice, and not as required by parents and an authoritarian culture, as was the case in the rest of the world.

With equality and freedom came another virtue; innovation. As we shall discover in later chapters, these three are always linked and are dependent on one another.

***

Of watches, watermills and spectacles.

A lot was happening in the rest of the world. Great inventions such as gunpowder, paper, mathematics and advances in navigation were emerging. In India, the Taj Mahal was being built, one of the most beautiful of all buildings. In the city-states of Asia and the Americas, they built pyramids and grand monuments. However, they were still ruled by civilisations based on domination and poverty. Meanwhile, the West took the best of the foreign inventions – and improved on them.

The following is taken from the Documentary. Genius of Western Civilisation

> "The alphabet, commonly used worldwide today, came originally from the Phoenicians. It was improved and adapted. The system of numbers, from zero to ten and so on, came from India through the Islamic empires of Asia and Africa. Again, it was improved and adapted. In Britain, the Romans left behind a network of water mills, some 6000 of them, designed to mechanically grind flour from grain. This design was improved and adapted so that when Chinese hand-made paper appeared in the West, Britain turned paper-making into a mass production using their water mills."

Man has always been intrigued by the passage of time. Sundials were first used in Egypt around 1200 BC and later by the Babylonians, Greeks and Chinese. The Chinese invented incense clocks. Islamic water clocks were unrivalled in their sophistication until the mid-14th century. In Europe, the demand increased for ways not only to tell the time when the sun wasn't shining but also to accurately measure the passage of time between two points. Many of these creations were mechanical monstrosities, such as the early town clocks boasted of by all leading towns and villages in Europe.

Then, on the back of discoveries in metallurgy and mechanical devices such as lathes, the humble clock morphed into the watch. A highly mechanised, extremely accurate timepiece composed of minute cogs, gears, springs and eternally whirring wheels. In the 16th century, in Germany, mass-produced watches became so compact they could be carried in a pocket or strapped to a wrist. Soon, clocks of supreme accuracy and reliability went to sea. The chronometer, an extremely accurate clock designed primarily by the Swiss, revolutionised navigation and was used for celestial measurements and determination of longitude.

So now we have a written language to record, a numerical system to quantify, the capacity to accurately measure periods of time, paper to publish and the invention of the Guttenberg printing press in 1440. Finally, the Bible, the most popular of all books, is in the hands of the common people. And now, we have all the necessary ingredients for modern Science to emerge and flourish. Except for one, perhaps the most important of all.

After a three-year wait, the cataract that had made me blind in one eye was finally fixed this week by a marvellous operation, inserting a new lens, all due to the wonder of modern science and medicine.

One of the almost universal curses of middle age is short-sightedness. Even with innovations such as oil lamps and other lighting that encouraged reading long into the night, it did not help those who could not see. In the late 1200s, an Italian monk contemplated the qualities and shape of a lentil when it occurred to him that a clear lens of that same shape could be used to redirect light and magnify eyesight. Although the invention of lenses can be traced to ancient times, here is another example of innovation and improvement. He made a pair of identical lenses that, when joined together, could be perched on the nose to improve shortsightedness. Suddenly, millions of middle-aged people across the Western world would continue to read and study. Suddenly, they became more productive. This was a game-changer.

✻ ✻ ✻

Men of God and Science.

The premise of this book is that Science and The Bible can co-exist, often agree, and occasionally disagree. But beware. There is a distinction to be made between the Bible and The Church. They are not the same. Many of their differences were profound, sometimes leading to bloodshed, wars and death, and invariably settled by a re-interpretation of the facts. For instance, for over two thousand years, both Science and Church believed that the universe was eternal. That there was no beginning and no ending. This was contrary to the Bible, which clearly states there was a beginning, in the beginning! It was not until 1965, only a few decades ago, that Science was forced to admit error and regroup. This is what happened.

From an article by The European Space Agency: Cosmic Microwave Background (CMB) Radiation

> "Penzias and Wilson, two radio astronomers in the United States, registered a signal in their radio telescope that could not be attributed to any precise source in the sky. It apparently came from everywhere with the same intensity, day or night, summer or winter."

It soon became apparent that this was a remnant of radiation waves from the Big Bang, an echo from the universe's beginning. In this case, Science and the Bible now agree. In fact, the Bible had clearly stated that there was a beginning for several thousand years.

Dr Stepen Meyer agrees:

> "In addition, many historians of Science have shown that belief in God served both as an inspiration for doing Science and as a framework for explaining scientific observations during the crucial period known as the scientific revolution (roughly between 1500 and 1750), in which modern Science as a systematic endeavour first originated.
>
> The founders of modern Science assumed that if they studied Nature carefully, it would reveal its secrets. Their confidence in this assumption was grounded in Greek and Judeo-Christian ideas that the universe is an orderly system—a cosmos, not a chaos."

So, let's take a closer look at some of the giants of the scientific method and experience what they thought and believed.

Many of those who studied Nature in the medieval world, which spawned the modern scientific method, were very materialistic in outlook and worldview. They believed that the natural world and everything in it are composed only of matter. That it is a purely mechanical world comprised only of material things. Even the mind and consciousness are claimed to derive from matter. This view is as common today as it was back then.

Johannes Kepler disagreed. He was a German [astronomer](), [mathematician](), [astrologer](), [natural philosopher]() and lover of music. A key figure in the 17th-century Scientific Revolution known for his laws of planetary motion and many books, he influenced Isaac Newton in his later work on universal gravitation. The variety and impact of his work made Kepler one of the founders and fathers of modern astronomy, the scientific method, and modern Science.

As a devout Christian and Bible scholar, he wrote a letter dated 10 Apr 1599 to the Bavarian chancellor Herwart von Hohenburg

> "Those laws of Nature are within the grasp of the human mind; God wanted us to recognise them by creating us after his own image so that we could share in his own thoughts."

In a personal letter, Kepler was also an advocate for Church Unity, for example:

> …. arguing that Catholics and Lutherans should be able to take communion together. He wrote, "Christ the Lord neither was nor is Lutheran, nor Calvinist, nor Papist.

Kepler died age 58, in Regensburg, Germany. However, he is remembered for his laws of planetary motion and NASA's Kepler space telescope, named after the great man, which discovered thousands of exoplanets between 2009 and 2018.

…..

It was not long before the technology for making clocks and watches was used to create models demonstrating the passage of the Earth's sun and planets. It was believed that all travelled in perfect circular cycles, the Earth at the centre, surrounded by the sun and planets.

Since the seventeenth century, the Italian Galileo Galilei has been regarded as one of the true fathers of modern Science. He is one of the earlier inventors of the telescope, where a lens at each end of a tube could magnify the image viewed many times over. He was the first to report telescopic observations of the mountains on the moon,

Jupiter's moons, the phases of Venus and the rings of Saturn. And this was over five hundred years ago.

In The Return of the God Hypothesis, Stephen C Meyer continues:

> "Galileo is often associated with conflict between faith and Science. The true story is both more complicated and more interesting. One of the true turning points in the history of Science occurred 22 years before Galileo was born. A work by Nicolaus Copernicus, published in 1542, proposed a heliocentric system in which the Earth revolved around the sun. For centuries, the Western world thought that humans stood at the centre of the universe. It was the system inherited from Ptolemy and Aristotle. They believed the Bible, and scores of renowned interpreters of the Bible, all seemed to point toward an Earth-centered or geocentric universe."

Thus began the famous conflict between Galileo and The Church that would last for the remainder of his life. Although he was a practising, Bible-believing Christian, he believed the Catholic Church's interpretation of the Bible to be incorrect. He challenged it through his many books and papers, gaining popular support amongst the general public and scientific community. However, the Church refused to have its authority challenged. Finally, it convened an Inquisition of Bishops and Cardinals to deal with the heresy. They claimed that the Earth was the centre of the universe, surrounded by planets and the sun. They quoted scriptures that Galileo claimed were meant to be understood figuratively. They disagreed, and he was forced to recant under penalty of death! Note: the interpretation was in error, not the scriptures!

So, the most outstanding scientist and thinker of his time was condemned to house arrest for the remainder of his life.

***

Galileo Galilei died alone in his own house on 25th December 1642. It was the end of an era.

Or was it? On that same Christmas day, 1642, a tiny, sickly boy child was born on a lonely farm in Lincolnshire, England. Someone said this was God's way of ensuring at least one genius was always on the planet! His father, an illiterate farmer, had died a few months before him. An inauspicious start for a man who would one day be buried in Westminster Abbey, amongst Kings and Queens and reverently attended by Nobles, Lords and Princes of the Realm! His name was Isaac Newton. Perhaps the last of the "natural philosophers" and the first of the great scientists.

When asked which scientist he'd like to meet, Neil deGrasse Tyson said, "Isaac Newton. No question about it. The smartest man ever to walk the face of the Earth. The man was connected to the universe in spooky ways." Watch live here.

He was so small his mother said he could fit into a quart jug. When he was three, she remarried and left Isaac in the care of his maternal grandparents. His early years were modest, and although his mother was widowed once more seven years later, Isaac never forgave her for abandoning him. A trait of unforgiveness that would characterise his later life. He would always have difficulty handling disagreements and criticism.

An unremarkable childhood, in which he neither excelled academically nor failed, saw him attend the King's School from age 12 to 17. He was competent in Latin and Ancient Greek and probably had a foundation in Mathematics without any significant signs of aptitude. In 1661, he attended the University of Cambridge. Although a scholarship eluded him for several years, he became a valet and servant to the senior boys. But then came what he later described "as the most inventive days of my life." Like COVID, the plague forced him to spend two years back at his country home, where he lodged with a prominent apothecary, like a modern-day chemist. He must have had a considerable effect on the young man, as he would become an ardent alchemist in later years.

During this enforced isolation, something clicked in the young man's mind. It was as if God had infused his ordinary mind with genius elements! As Johannes Kepler himself said:

> "God wanted us to recognise natural laws and that God made this possible by creating us after his own image so that we could share in his own thoughts."

He used the break to develop his ideas of gravitation, optics, and calculus, which he used to write his masterpiece Principia, which is so familiar to modern university students of all persuasions. Those who sought a materialistic or mechanical explanation widely disputed his original thoughts on gravitation as a force. But it endures to this day, becoming somewhat modified by Einstein's Laws of Relativity. Because he delayed the publication for some years, he was challenged over the original discovery of Calculus. Still, he overcame that after many years of often bitter arguments.

He never married and was by all accounts a dour, serious man, quick to take offence, with as many enemies as friends. In a long and spectacular career, he was President of the prestigious Royal Society for many years, was twice elected to Parliament, and was appointed Master of the Royal Mint for the last thirty years of his life. During this time, he told the now-famous story of how an apple falling from a tree influenced his ideas on gravitational force.

A prodigious writer of scientific texts, he later turned to the Bible, mainly Church History, Prophecy and Theology. Relying on the books of Jeremiah and Daniel, he accurately predicted the return of the Jews to Israel in 1895 and 1945, a remarkable feat for a 17th-century scholar. He wrote far more on theology and alchemy than he did on Science, with millions of words on these subjects alone. He was a serious Bible scholar and, at times, questioned the doctrine of the Trinity. However, he was careful to keep this quiet during his years of scientific discovery.

These religious papers have come to light recently and finally found a home in the Israel Museum. More recently, they have been extensively catalogued by a team at Oxford University. The following is taken from the Documentary. Genius of Western Civilisation

> Since January 2008, The Project has published over four million words of diplomatically transcribed text, about 90% of which is made up of Newton's technically forbidding writings in physics, mathematics and theology. These are foundational documents in Western and global intellectual history, and they are published in full, with accompanying images of the originals, for the first time.

Sir Isaac Newton is regarded today as one of the most outstanding scientists of any age. Yet, like so many of his fellow scholars in times past, he was a devout and committed Christian who found little conflict between his beloved Bible and his Science. He shared this viewpoint with many of this era: scientists such as Bacon, Descartes, Kepler, Christiaan Huygens, Richard Hooker, and Sir Robert Boyle. Modern scientists stand on the shoulders of these giants.

***

Two ways to experience Nature.

So far, we have looked at the Bible as a historical document and the impact of the Judeo-Christian faiths on Western Civilisation, leading to the birth of the modern scientific method. We studied a handful among many of the early natural philosophers as they struggled to interpret Nature in the context of their Biblical understanding.

We can use this understanding to explore the natural world, Creation, in all its glory and wonder.

In The Return of the God Hypothesis, Dr Stephen Meyer continues:

> "Early in the Christian era, theologians began referring to Nature as a book, one that they likened to the Bible in its ability to reveal the attributes of God. Just as the book of scripture told of God's character and plan, so too did the book of Nature reveal God's power and wisdom. As early as the third century, the Christian monastic Anthony the Abbot referred to "created nature" as a "book," one always at his "disposal" whenever he wanted "to read God's words.
>
> Viewing the natural world as a book that reveals the character and Nature of God provided a theological inspiration for the formal study of the natural world. It also reinforced conviction in the intelligibility of Nature because it implied that the divine

author not only speaks through the book of Nature but that men and women made in his image and endowed with his rationality were equipped to read and understand it."

Indeed, the Bible itself declares how we should study Nature in Psalm 19 1:4:

> *"The heavens declare the glory of God;*
> *the skies proclaim the work of his hands.*
> *Day after day they pour forth speech;*
> *night after night they reveal knowledge.*
> *They have no speech; they use no words;*
> *no sound is heard from them.*
> *Yet their voice goes out into all the Earths*
> *their words to the ends of the world."*

However, as Johannes Kepler has already pointed out, Nature can also be understood as a set of laws. The Laws of Nature.

> "God wanted us to recognise natural laws and that God made this possible by creating us after his own image so that we could share in his own thoughts."

The early fathers of modern Science also quickly adopted the analogy of a clock and its mechanism to illustrate how God works through His Creation.

> "Numerous historians and philosophers of Science and a few scientists have made this connection. Some have even identified the Hebrew Bible as the ultimate source of the metaphor. As the Nobel laureate and University of California–Berkeley chemist Melvin Calvin argued, the notion of an

"Order of Nature" was "discovered 2,000 or 3,000 years ago and enunciated first in the Western world by the ancient Hebrews." Calvin notes that the monotheistic worldview of the ancient Hebrews suggested a reason to expect a single coherent order in Nature and thus a single, universally applicable set of laws governing the natural world.

Next, we will study the three significant issues that continue to perplex modern Science and many Christians. The Creation of all things, the emergence of life, and the beginning of human consciousness.

We will refer to the Bible, Modern Science, and ancient Hebrew texts through the prism of the Book of Nature and the Laws of Nature. Once again, the duality of reality!

It promises to be an adventurous and spirited journey.

Summary:

Science and the Bible are complementary works, with the Bible ascendant.

Nature should be read as a book, revealing God's word and plans for creation - the duality of reality.

Modern science was born in the crucible of Western civilisation.

Contrary to popular belief, most early scientists were Bible believing

CHAPTER 3

Beginnings

Humans are awed as the Milky Way galaxy stretches over our upraised heads like a magnificent glowing canopy. This galaxy is our home. It is populated by over 100 billion stars like our sun, a number so vast as to be almost unintelligible to human understanding. Furthermore, there are 200 billion other galaxies like ours in the whole universe. Yet amazingly, there are about the same number of connecting neurons in the human brain! Inside your own head. A system far more complex, problematic, and worthy of study and attention than the greatest of those cosmic wonders. We are about to become far more interested in the <u>mind</u> of Creation than the physical Creation.

In this chapter, we will explore our understanding of the very beginning. Why? Where? How? We'll briefly study the physical Creation, first called the Big Bang by a sceptical Sir Fred Hoyle:

> Fred Hoyle, a renowned English astronomer, was initially a staunch advocate of the steady-state theory of the universe. He proposed that the universe was eternal and unchanging, contrary to the popular Big Bang theory. However, in a strange twist, Hoyle inadvertently coined the term "Big Bang" during a radio interview in 1949 while criticizing the theory, and the term stuck.

Like most scientists of his day, and for thousands of years prior, Hoyle derided any suggestion of a beginning for the universe. In this, he and all the others stood directly against the Bible, which declares a beginning in its opening statement. Later in life, he supported the Big Bang theory as evidence grew. So, both he and the Bible agreed: there was a beginning.

In the prologue to my recent book Makarrata, I described the beginning like this.

> Consider for a moment, if you will, the concept of – nothing.
>
> No people nor parents, no cousins or friends, no life at all. Just nothing.
>
> No land nor sea, no birds that fly, no sky. Just nothing.
>
> No earth nor sun. No stars that shine. No moon. Just nothing.
>
> No light nor dark. No space. Not even a vacuum. Just nothing.
>
> No, not even time. Time requires matter, and matter requires energy — just nothing.
>
> Complicated. Perhaps it is impossible for humans, to understand, bound inexorably to this glorious planet on which we live and breathe? It would be unnatural to imagine what we have not experienced. With all its innate intelligence, with the human soul, consciousness is incapable of conjuring up – just nothing. Our mind needs, nay demands, something.
>
> But wait. We cannot experience just nothing. We cannot imagine it with our minds. But there is a way. If we use not our consciousness but our Spirit.
>
> There. Look. That's it! There it is!
>
> Before time and matter, in the Spirit, we can perceive this. And yes, there is something. There is Spirit. An abundant,

overflowing, overwhelming intelligence. Wisdom is there. Can you not see it, feel it in your Spirit? Has it not been spoken of in awed, hushed tones around ancient campfires and throughout history past.

Oh, and there is something else. What we now call "The Laws of Nature." The immutable system of checks and balances that guides all Creation. Everything in the Universe. Those Laws are <u>already</u> there. Before anything was, they existed. Expectant. Waiting through eternity for the moment.

Michael Chambers. [Makarrata. The Australians of Arnhem Land.](#)

The Primacy of the Bible

I would encourage you to visit the [NASA website and search for the WMAP image](#) of the generally accepted scientific timeline of the universe. In graphic detail, we see both the beginning stages of expansion and our current position in space and time. On the far-left-hand side is the beginning, what *they* call a quantum fluctuation, the Big Bang, and what *we* call Creation. All the galaxies, stars, and matter that there ever was or ever will be come from something tiny! Including the 100 billion neuron connections in your brain!

We have established that the study of the natural world, that discipline we call now Science, is ancient and dates back to the beginning of mankind.

So, too, the Bible. It is an ancient book written by ancient people, for ancient people. It has endured many forms and translations. So, one would think that the closer we go to the source, the more we can get a sense of the original meaning. It seems reasonable for us to discover what the early scholars of the Torah thought and wrote about. Nahmanides was a Jewish Torah Scholar living nearly 800 years ago. Gerald Schroeder describes his ancient writings like this in his book, The Science of God.

> "Nahmanides ... described the process with uncanny accuracy: the initial Creation produced an entity so thin it had no substance. It was the only physical Creation ever to occur and was all concentrated within the speck of space that was the entire universe just following its Creation. As the universe expanded from the size of that initial minuscule space, the primordial substance-less substance changed into matter as we know it. Biblical time, he continued, starts ("grabs hold," in his words) with the appearance of matter."

Physics, during the past twenty years, has come to agree with him. Nahmanides writes that his teachers learned this account from the first word of the Bible, Be'rai'sheet, which means "In the beginning of." In the beginning of what, they asked? In the beginning of time was their conclusion. In the beginning of time.

Every ancient commentary, with no exception, tells us that the six days of Genesis were all six 24-hour periods no longer than the six days of our work week. Moses said in Deuteronomy 32:7

"Remember the days of yore; consider the generations long past."

In 1750 Vilna Gaon explained:

"To understand "the years of every generation," one must first remember the days of yore – the Six Days of Creation. For in those days lies the complete plan of the development of the universe and humankind in it."

This, he taught, is the only way to understand history.

In all of the Bible, in all of history, there has probably never been a more contentious statement than that made in the first 31 verses of Genesis. God created the world in six days and rested on the seventh. It is a controversy that continues today. In the past, differences over this have led to violence and death. But, if we had listened to the early commentators and scholars, the correct interpretation would be that *both* 14 billion years *and* six 24-hour days are accurate. The reality of duality! In the next chapter, we will learn how this can be true.

Doctor John Lennox is an Irish scientist and mathematician holding four doctorate degrees and the prestigious Chair of Mathematics at Cambridge University. In his book, "Seven Days that Divide the World," he proposes this about the Bible and Science.

> "It is Scripture that has the final authority, not our understanding of it.... Even though our interpretation relies on scientific knowledge, it does not compromise the authority of Scripture. And this is the critical point. Scripture has the primary authority. Experience of the world in general and Science, in particular, has helped decide among the possible interpretations that Scripture allows."

Maimonides known as the Jewish Ramban, who died in 1204, said: "Science is one of the primary paths to knowing God, so the Bible commences with a description of the Creation."

In the twentieth century, Albert Einstein (1879 – 1955) changed the course of Science forever with his Theory of Relativity. He said: "Science without religion is lame. Religion without Science is blind. If God had created the world, His primary concern would certainly not have been to make its understanding easy for us. Before God, we are all equally wise and equally foolish."

Summary:

The Bible has primacy over everything. But it is scripture, not our own interpretation.

Modern science and Torah both agree that there was a beginning. The ongoing debate is about who or what began it!

CHAPTER 4

Seven Days that Divide the World

Have you ever seen one of those diagrams designed to trick you into what you are seeing? I'm sure you have. My favourite shows the outline of two opposing faces. But then you blink – and lo, it becomes a candlestick! And there are many others. However, the reality is that there is only one diagram that depicts two different images. It is the value of perspective. To see things from a different viewpoint. And that is what we are going to discover here.

We have established that the Bible is the preeminent source. The final judge. So first, we will discover what modern Science says about the universe's beginning. Then, we will compare that to the biblical perspective.

This chapter will reveal how the universe expanded 6 days into 13.77 billion years. The following chapters will study the 31 verses of Genesis to determine what the 3500-year-old Torah says. And hang on – this could get exciting!

The Torah looks at the universe from a universal perspective. It looks forward from the Big Bang and sees the 6 days of Creation. It is as though a narrator is describing what happened as it happens. Now, <u>this is important.</u> The days <u>cannot be</u> from an Earth-based perspective as the Earth was <u>not created</u> until day 3, as we shall discover later.

In the beginning, according to Science, there was an event. It is not really known what that event was or what caused it. It is best described as a "Singularity" or a "Quantum Fluctuation." (Someone once quipped that if you have studied Quantum Physics and are not confused, then you haven't studied Quantum Physics!) There was no bang because there was no atmosphere to conduct sound. There was no explosion, where matter and debris were flung into space in all directions. It was an immediate and all-embracing expansion – inflation- that continues to this day, billions of years later. Hard to imagine? Try this.

You are making a pudding. You need yeast to bring it to life. You roll and knead the dough into a small ball. It is full of sultanas, all jostling together inside the dough ball. But look what happens, first when it rises and then when you put it in the oven. It expands rapidly at first and then more slowly over time. As the size increases, each sultana moves further away from the others. Expansion. Inflation.

***

Next, matter starts to form; time grabs hold! The beginning of time. From this perspective, the Torah describes the following events from the beginning of time. Within a relatively short period, about 380,000 years, cosmologists say, the first light appears, more of a glow than a beam. The process of building the universe continues for aeons. Stars continually form and explode, sharing their stardust across the universe and reforming into stars and planets. Nearly 14 billion years later, the James Webb telescope, hovering in space a million miles from Earth, receives packages of light in the form of photons that have been travelling for billions of light years!

According to the leading textbook on the subject, Peebles. Principles of Physical Cosmology.

"The expansion of space stretches time by the same amount that the wavelengths of light travelling in space are stretched. Time is stretched by the expansion of space by the same amount that light waves travelling in space are stretched or red-shifted."

Light always travels at a constant speed. The speed of light is 299,792,458 metres per second, so we expect the wavelength to be constant. Yet the stretching of space causes light waves to be stretched or red-shifted as they travel through space. This is called Cosmological Time Dilation. The stretching of time is caused by the expansion of the universe.

Here is another way of looking at it.

My son owns a Virtual Reality arcade. Imagine you and two friends, all the same age, go there to experience one of the machines. The Time Machine. You don your headsets and begin in a galaxy far, far way! Then, a second apart, you each jump on a light beam heading in the general direction of what will one day be Planet Earth. From first to last, you are separated by two seconds. But look what is happening. The fabric of space itself is stretching while you travel. And it is stretching time as well. So, when your three light beams reach the James Webb telescope, you are no longer two seconds apart but ten seconds apart. Space-time has stretched. Oh, and one other thing. At the speed of light, time ceases to flow. So, despite the length of your voyage, all three of you are the same age as when you left!

In his book The Science of God, Dr Gerald Schoeder describes the expansion of time and the universe like this:

"Light, you see, is outside of time, a fact of nature proven in thousands of experiments at hundreds of universities. Light is outside time. Light, existing outside of time and space, is the metaphysical link between the timeless eternity that preceded our universe and the world of time, space and matter within which we live. As with all light-like radiations (the photons of gamma rays, X-rays, light, microwaves, etc.), light can abandon the ethereal, timeless realm of energy and become matter. In doing so, it enters the domain of time and space. Einstein's famous formula, $E = mc^2$, teaches that light and matter are two forms of the same thing. In terms of days and years and millennia, this stretching of the cosmic perception of time by a factor of a million million, the division of fourteen billion years by a million million reduces those fourteen billion years to six days!"

What was once six days viewed from Creation is now 13.7 billion years viewed from Earth! And yes, the maths does stack up, but it's a bit complicated to show here. Mitch Goldberg from the Headbangers YouTube Channel has a presentation based on Dr Gerald Schroeder's writings, Genesis and the Big Bang. His excellent presentation clearly demonstrates the cosmology and mathematical formulae proving this theory. I highly recommend you seek it out.

Depending on your perspective and direction, the six days of Genesis and the nearly 14-billion-year-old view of the universe are the same.

Here is how Mitch Goldberg summarises this:

"If the Torah is a man-made fiction like every other creation story, then Moses could have made anything happen for his creation fairy tale. But he described this sequence of events over six days. How could Moses have known this over 3500 years ago? Most of what we know was only discovered in the last few hundred years.

This is another indication that the Torah could not have been authored by a human but was indeed inspired by God.

Torah's 3500-year-old description of Creation in 6 days is scientifically accurate, This demonstrates the incredible truth of the Torah. In Genesis and the Big Bang, On YouTube, Mitch Goldberg explains:

> Some 250 years ago, a great sage and mystic, the Gaon of Vilna, taught that when the light of Torah came into the world, it split into two parts. Only one part was revealed directly, the prophetic experience. The other part was hidden in the wisdom of nature, and the time will come, he said, when those hidden wisdoms of nature will be discovered, revealing aspects of the Torah never before understood. The hidden wisdoms of nature are being discovered. Einstein's relativity, the secrets of light, and the atom all revealed a new insight into the six days of Creation.

***

I have tried to keep complex theories and calculations out of this explanation, so bear with me as we go through this. They are readily available elsewhere. Torah describes what happens each day. In today's terms, we can use the same equation to calculate the number of days in the previous example to determine the approximate length of each day. This is the result.

Day 1 began 14 billion years ago

Day 2 began 8 billion years ago

Day 3 began 3.8 billion years ago

Day 4 began 1.8 billion years ago

Day 5 began 750 million years ago

Day 6 began 250 million years ago

If we were to look at those days mathematically, Genesis time and Earth time, it would describe something like a cone, with Day 1, at the pointy end of the cone, the longest day, progressing to Day 6, the largest end of the cone, the shortest day. This can be described as an exponential curve. The cone is a typical example of this pattern in nature, as Dr Gerald Schroeder explains in The Science of God:

> "The graceful curve of the Nautilus seashell occurs in nature more often than any other shape. Its lines trace out an exponential spiral. Each of the spiral's successive swirls is wider by a seemingly arbitrary but fixed factor. We see the curve repeated in the distribution of seeds on a sunflower, the curves of tusks, and the spread of stars in spiral galaxies, our own Milky Way included. Graphically, it also describes the relation between Genesis time and Earth time as the universe expanded from its point-like creation of the big bang."

Summary:

The Bible is silent about the length of time between Creation and Adam, except that it differentiates between the time leading up to Adam being regarded differently.

However, we have established in a methodical scientific way how six 24-hour Earth days can also be viewed as billions of years by the stretching of time caused by the universe's expansion. The duality of reality.

This satisfies both our scientific understanding of the natural world and the theology of the Bible.

CHAPTER 5

Genesis Day One

Before we look at the 6 days of Creation in Genesis, we should remember the circumstances in which the Torah was written. Reflect on the story at the beginning of this book. In the ancient bronze age, people created the art of writing to record and inform their own population of events that continue to amaze even our sophisticated modern minds. Moses did not write a scientific textbook; he told a story. But even after 3500 years, the story remains essential to our understanding of the world.

For the past couple of years, I struggled with a cataract in my left eye. It deteriorated so that, for all intents and purposes, I was partially blind. I could still see to read and write with one eye. With only one eye, judging close distances, especially when pouring from a jug into a glass, is severely impaired. Even so, my surgeon assured me that after the operation, with two good eyes, I would return to near-perfect sight.

If we only look at this story through one eye, we will miss the richness and depth of the narrative. Let's look at this story with two good eyes, the Biblical and the Scientific. We will discover far more richness and revelation. Neither view contains all the truth – but much more is revealed when viewed together.

In the previous chapter, we have seen how a scientist gives approximate dates to the age of the universe and calculates the age and length of the six biblical days of Creation. These can only be approximate. The calculations and variations can change from year to year. And that is to be expected. But the Torah has been constant for more than three thousand five hundred years. That also is to be expected.

Science says the First Day started 13.7 billion years ago.

In Genesis 1:1-5 Complete Jewish Bible says:

> "In the beginning, God created the heavens and the earth.
>
> The earth was unformed and void, darkness was on the face of the deep, and the Spirit of God hovered over the surface of the water.
>
> Then God said, "Let there be light" and there was light.
>
> God saw that the light was good, and God divided the light from the darkness.
>
> God called the light Day and the darkness he called Night. So there was evening, and there was morning, one day."

As I describe these events in my recent book, Makarrata:

> Everything that ever was or will be is there in that beginning. Every particle, every atom of our being, and everything around us is all there, everything. Every star that was born from every galaxy is there. We are truly made of stardust.
>
> It is a state so crammed with energy and heat that matter cannot form. However, as we know, $E=mc^2$, energy is matter, and time will measure the response.
>
> Slowly, inevitably, there is cooling. Soon, the temperature falls sufficiently for matter to form. And it does – and billions upon

billions of particles form. And so, it would have progressed. An infinite, perfectly formed and balanced universe, leading – nowhere. For without matter, the great cosmos lies bereft of life. And this current, perfectly formed state will never lead to life. It will be dormant, boringly inert. Life-less.

But now there is another "fluctuation." Some might describe it as an intervention. The perfect becomes imperfect, from lifeless to a future designed for life.

The incredible universe ceases its sterile existence and becomes a matter-making factory. Protons and neutrons now combine with electrons. Atoms are formed. Hydrogen is one of the first. So, when your body, mostly water, next enjoys a glass of that miraculous liquid, ponder this; the hydrogen atoms you drink were there at the beginning!

Time passes, as it is now able to do. The first four hundred thousand years pass. Creation is now the size of the Milky Way. It consists of neither liquid nor solid but rather plasma, dense and opaque. There is more cooling. Suddenly, for the first time, there is light. Not blinding, bursting, shattering light, but a soft glow emerges. And grows. So, for the first time, there is darkness, and there is light.

Makarrata: The Australians of Arnhem Land

Be'rai'sheet (Bra-sheet)

Dr Gerald Schroeder is a Jewish physicist who has worked extensively with NASA and the Atomic Energy Commission. He also holds a PhD in Earth and Planetary Sciences and Nuclear Physics from Massachusetts Institute of Technology. He says: "I am just the lucky guy who knew enough Torah and science to put the numbers into equations."

He spent 7 years on the staff of the MIT Physics Department. Before moving to Israel, he joined the Weizmann Institute of Science, the Volcani Research Institute, and the Hebrew University Isotope Separation Mass Spectrometer facility. He has formal theological training and over 20 years of study under the late Rabbi Herman Pollack, Rabbi Chaim Brovender, and Rabbi Noah Weinberg. In other words – he is a smart guy!

He is also the author of four best-selling books. The Science of God, Genesis and the Big Bang, God According to God, and The Hidden Face of God. He is a guest on numerous podcasts and interviews. He has also published many scientific papers that he has generously allowed me to use and quote in this book. In The Science of God, he says that it is generally understood among the scientific community that the Big Bang also marks the beginnings of The Laws of Nature.

> "…. prior to the existence of the universe, there was no nature and therefore, there were no laws of quantum mechanics by which to engender the needed quantum fluctuation. …… Alternatively, we might believe that the laws of nature are eternal. That would suppose the existence of laws for a universe that does not exist. Purposeless laws. This stretches one's imagination. If on the other hand we accept these eternal laws to be part of something grander, we are back to the Bible. Theology claims the laws of nature are indeed eternal."

The Bible teaches that God is eternal. That His Creation was premeditated, a deliberate action with an end result in mind.

According to the Bible, the endgame was life on earth. Consider the uniqueness of life on that lonely, isolated planet orbiting an average nondescript star, our sun, in the corner of one of the smaller galaxies. So, how did He accomplish that Creation? In John 1:1, according to the Apostle John, who writes in the New Testament - He spoke!

> "In the beginning was the Word, and the Word was with God, and the Word was God."

So, get this! Our world and everything living in it, all the stars and galaxies, the whole of Creation, came about through wisdom and a spoken word. In fact, we will learn in the following chapters that the physical universe, which we consider fundamental, is an underlying substance of a creation made from mind, wisdom, and intelligence. How do we know? Because the Bible tells us so! In "God According to God" Gerald Schroeder says:

> Everything from our bodies to boulders on a mountain is made of the energy of the Big-Bang Creation. The scientific discoveries of the twentieth and twenty-first centuries have gone a step farther in closing ranks with the Creation, finding that matter and the energy from which matter is formed are made of something totally ethereal. In physics, we call it information or, more extreme - mind. In the words of the knighted mathematician James Jeans, the world looks more like a great thought than a great machine. Biblical theology agrees, telling us, as we will learn, that God used a substrate of wisdom to build the world. This Divine wisdom or mind is present in every iota of the world's being. It explains how the energy of the Creation, essentially super powerful light beams, could become alive and

conscious, able to feel love, joy, and wonder. Divine wisdom was and is present, guiding and forming the way.

And what was that first cause? The twenty-one-hundred-year-old Jerusalem translation of the Bible into Aramaic, a sister language of Hebrew, holds the answer. B'raisheet is a compound word meaning "with" or "using" or "a first cause" (raisheet); hence, "With a first cause [of wisdom] God created the heavens and the earth." That the "first cause" is defined as wisdom arises from Proverbs: "I am wisdom…. God acquired me [wisdom] as the beginning of His way.

Gerald L Schroeder.. God According to God

The Bible describes the Creation of the Universe as a series of speech events.

As Gerald Schroeder describes it, the Jerusalem translation of the Hebrew Bible, the Torah, is one of the oldest translations. It translates the original Hebrew into Aramaic, a language commonly found in the Old and New Testaments. "…the opening sentence (Gen. 1:1) should be understood to mean, "With a first cause God created the heavens and the earth." That first cause was then defined as wisdom. This was derived from the book of Proverbs chapter 8:

"I am wisdom…God made me [wisdom] as the beginnings of His ways,

the first of His works of old. I [wisdom] was set up from everlasting,

before ever, there was an earth.

When there were no depths, I [wisdom] was brought forth.

From here, we learn that the "first cause" to which Genesis 1:1 refers is wisdom, and with this insight, we read the opening sentence of the Bible as

"With wisdom, God created the heavens and the earth."

The whole universe was created, not with a bang, but with a spoken word. A language containing all the wisdom and intelligence necessary for future Creation. Hebrews 11:1 says:

"In the beginning was the Word."

We might say that Creation was "word-based!"

In "The Science of God," Dr Gerald Schroeder describes it well:

> ... but the idea that information is primary is central to modern physics and biology. More importantly, since it is word-based, the universe still has meaning.
>
> Here, then, is the Genesis Enigma: The opening page of Genesis is scientifically accurate but was written long before the science was known. How did the writer of this page come to write this creation account?
>
> Surpassing the harmony of the laws of physics and the workings of biology lies the subtle truth that every bit of existence is composed of a single substrate—energy—created at the beginning. Einstein theorised, and a multitude of experiments have since tested the truth of this bizarre and totally illogical reality.
>
> Gerald L Schroeder.

Summary.

The opening passage of the Bible makes the most astonishing claim in the history of humanity. It describes the first and foremost of many miracles, as God enters his Creation- with intent.

Consider this: let's say you can simply commit to believing that, with wisdom, God created the heavens and the earth. If that is so, then the rest of the Bible and all the miracles it describes in great detail, including the resurrection of Jesus, are not only believable but entirely understandable.

Paul wrote over 2000 years ago in his letter to the Hebrews: "Now faith is confidence in what we hope for and assurance about what we do not see."

CHAPTER 6

Genesis Day 2

The second day started 8 billion years ago.

The Bible says:

God said, "Let there be a dome in the middle of the water; let it divide the water from the water."

God made the dome and divided the water under the dome from the water above the dome; that is how it was,

and God called the dome Sky. So, there was evening, and there was morning, a second day.

Genesis 1:6-8 CJB

Except at the nuclear level, where quantum mechanics can alter statistical probability, the laws of physics that predict the formation of stars, galaxies, and elements rely on the occurrence of the probable over the improbable. The probable and improbable

About ten billion years pass, more or less. In a remote corner of the Universe, tucked away in a minor galaxy, well out of the mainstream of activity, there is a massive cloud, an object of considerable beauty. Composed of primordial dust, matter and particles collected from the debris of millions of dead stars and supernovas. All the materials needed to build our planet are now in place. Critical mass is reached. In the centre, gravity begins to exert itself, drawing matter tighter into itself. It starts to coagulate, like clotting blood. It begins to spin, slowly at first and then much faster and denser. The glowing core separates. It is huge and becomes so hot that fusion, a continuous nuclear reaction, begins, which will continue for billions of years into the future.

A star is born. It is our sun, the glorious life-giving centre and beginning of our solar system.

I continue the story in Makarrata:

> The matter left over eventually forms into about twenty planets. Over time, these begin to collide and bump into each other – generally forming one much larger planet. They become like vacuum cleaners, exerting gravity. Dominating, hoovering up most of the remaining debris and vagrant matter in our unique but separate region of the universe.
>
> The planet Earth, for thus it can now be known, a hot and fiery place, spins its way across the solar system on a course forever bound around the sun.
>
> As it rotates, the heaviest particles, clump, and objects are coerced to form a molten ball in the centre. Chief among the heavy elements is iron, fiery and molten. The radioactive composition of this central core will keep it hot for billions of years. The iron core also forms a vast unseen magnetic force field, protection that slows the attrition of the earth's formative atmosphere from the fierce solar winds that emanate from the sun. Other planets, such as Mars, will soon lose their atmosphere in this manner.
>
> The earth was cooling and roughly formed three and a half billion years ago. The formation of rocks begins. The ancient and oldest foundation materials are forced to the surface either

in a molten state, magma or as cooling rock, known as granite. Violent volcanic activity rules the surface, either as eruptions of vast flows of molten magma that, in turn, form a hard, rigid rocky material as it cools. Image a bowl of porridge bubbling away on the stovetop. It bubbles away freely for a while, but as time passes, the bubbling slows, and skin begins to form. Over time, this will create a crust like the one we have on the planet, which we call home.

At this time, the earth is formed internally, much as it is today. It is a globe, around six thousand four hundred kilometres from the centre to the poles. It is not perfect, though, because it is about fifty kilometres longer to the equator. The molten core rises about twelve hundred kilometres toward the surface. Around that is a cover, neither solid nor liquid, because nothing could liquefy at that pressure level. This is about two thousand kilometres thick. Around this again is a three-thousand-kilometre-thick mantle of dense rock, neither solid nor liquid but providing an elastic stretching buffer that protects and shields the surface from the most violent, fiery upheavals. On top of this lies the earth's crust, stretching barely forty-five kilometres below the surface and subject to the most violent eruptions and disturbances.

The baby earth is formed. At this time an inhospitable place with an atmosphere of mostly nitrogen, methane, and carbon dioxide. But two elements are missing that will be essential to future life. And both arrive while the planet is very young.

The debris from the newly formed solar system includes some minor rogue planets. In a macabre dance toward destiny, one such, about the size of Mars, stalks the Earth, growing ever closer and more dangerous. Finally, it strikes, and a glancing blow shatters the intruder into minuscule pieces of debris and matter that swing out into space like the crack of a giant whip. Over time, it clumps and forms into a tiny planet, tied by gravity to its rival forever. The Moon changes the future Earth forever, knocking it slightly off its axis to permanently have a wobble. This is where our seasons come from. In the future, it will also cause the tides to ebb and flow. How fortunate. (Some say that coincidence is God's way of staying anonymous!)

But for tides, we need water, and it arrives spectacularly. Far out in space, around Saturn, there is a massive ring of asteroids. It is not the usual space debris, but each contains large ice concentrations of frozen water. Many of these are knocked out of orbit by the wandering giant planet. As a result, the earth was bombarded by water carriers from the solar system's outer regions for half a million years. One only has to consider the Moon's pitted surface to realise how intense the bombardment is. So, the oceans arrive on the young planet from outer space, just in time for the Earth to bring forth life.

[Michael Chambers Makarrata. The Australians of Arnhem Land]().

We live on a planet made for life. As hostile as the frigid reaches of space may be, the vast universe provided the time and energy to allow stars to develop and die in supernova explosions, thus producing the elements needed for life and seeding them into space. Those same depths of space shield us from the lethal radiations from the supernovae.

The Luckiest Planet

A just-right Earth with just the needed gravity, radioactivity, magnetic field, and volcanic activity to support life is located at just the correct distance from the Sun to nurture the inception and development of life. Could this be accidental? But Earth should not be where it is. Among the planets circling the Sun, Earth is the oddball. The distribution of matter initially spiralling in toward a central attractor may reach an equilibrium that clusters along what is known as an exponential curve. In this curve, each successive swirl is a given factor farther out than its predecessor. Each planet is approximately two times farther from the Sun than the preceding planet, except for Earth. Earth should not be where it is. A just right earth. In just the right place and distance from the sun. Coincidence?

Summary:

The universe was designed for life from the moment of Creation.

As was Planet Earth, the luckiest planet in the universe.

The odds against life forming and flourishing on this tiny planet are so high as to be beyond calculation!

CHAPTER 7

Genesis Day 3

The third day started about 3.3 billion years ago

In Genesis 1:9-13 CJB says:

God said, "Let the water under the sky be gathered together into one place, and let dry land appear," and that is how it was.

God called the dry land Earth, the gathering together of the water he called Seas, and God saw that it was good.

God said, "Let the earth put forth grass, seed-producing plants, and fruit trees, each yielding its own kind of seed-bearing fruit, on the earth"; and that is how it was.

The earth brought forth grass, plants each yielding its own kind of seed, and trees each producing its own kind of seed-bearing fruit; and God saw that it was good.

So, there was evening, and there was morning, a third day.

The earth had cooled to allow the first liquid water to form, followed almost immediately by bacteria, first life and photosynthetic algae.

Oceans and dry land appear, followed immediately by life and then plants. The appearance of single-celled life so soon after the appearance of water astounds modern science for the speed with which it happened. There was no time for Darwinian evolutionary events to occur. That would take too long.

So far, on days 1 and 2, we have been looking at the emergence of the inorganic world. Material things. We have seen that when "God said," things happen. But then, on day 3, a change occurs. There is a second: "God said." And that triggers the origin of life, something that modern science has still not figured out. Note that God did not have to *create* life. God spoke to the planet. He commanded the Earth to bring forth the first life, with the elements already formed and at its disposal. However, God had to *specifically command* life into being. In his book Seven Days That Divide the World, John C. Lennox says:

> This could imply that God starts life going and allows its potential to develop from there. The point of the repetition of "And God said" in day 3 is to underline that life *did not* come from nonlife simply by processes of development. Life required not only pre-existent matter but also a special creative word of God.

It is essential to remember that science has not explained the two principal starting points in our lives: the start of our universe and life itself. The origins of life are a mystery to us. Despite many scientific claims to have solved the problem, creating life in a laboratory has proved extremely difficult.

So, here's the elephant in the room, Charles Darwin and his famous book "Origin of Species." Since its publication in 1859, it has swept the world by storm. Before Darwin, most thought life was inspired. Nearly a century later, evolution was considered by most of the scientific world and society in general to be the truth. It taught that the animal world came by means of millions of years and by continuing micro and macro evolution to produce the variety and diversity of life that we see today. Yet Darwin was largely silent on the origins of life. The next chapter will examine current scientific attitudes toward evolution and its consequences.

When I was growing up in the 1950s, it was widely taught in schools that life formed in some sort of nuclear soup, a premise widely held today, particularly in the broader community. We believed, as many do today, that experiments, if not conclusive, would soon lead to the absolute truth.

Dr James Tour says about the scientific understanding of the Origins of Life:

> That experiment has since been debunked! Nothing has changed in Origin of Life studies in 72 years. There is no road to life. There is the synthesis problem. Molecules have no brains; they don't keep records. They can't go back 400 million years as we can in a lab. They can't go forward towards a goal, because *they don't know where they're going*. Yet everything must be in its correct order or no result. There are known classes of chemicals necessary for life. Carbohydrates, Nucleic Acids, Amino Acids or Lipids.... The hardest thing is not to produce these acids and nucleotides but to hook up so that they work together. The cell has to be discerning what to let in otherwise it just dies. Many carbohydrates are so complex there are over 1 trillion ways they could be assembled. (That's $10^{12.)}$ If not assembled in the right order the cell dies. Mirror images of a compound are also necessary but very hard to make.

> Origin of life protocell assembly is akin to buying 20 pounds of sliced turkey meat, adding a gallon of turkey broth, warming, sticking in a few feathers, and suggesting that a live turkey will eventually appear if given enough time.

Here is an example that will help us understand these large numbers, million, billion, and trillion. Let's say you have asked your beloved to marry you. She says she will give you an answer in a million (10^6) seconds. That's 11 hours. That sounds reasonable! But what if she says she will give you an answer in a billion seconds. (10^9) That's 32 years. Not so good! But what if she says you will have your answer in a trillion seconds. (10^{12}) That's 32,000 years! Science says that's not impossible but unlikely. In this case, we would have to agree with science! It looks like she's not so keen after all!

Dr James Tour points out that these are the chances of life just creating itself. "Yet they tell us that on pre-biotic earth, from under a rock a fully formed prehistoric life form slithers out!"

Chances of that happening are $10^{79,000,000,000}$ There are only 10^{90} particles in the whole universe. Be amazed. These are big numbers!

Dr Tour continues,

> Critical for life is information and that is contained in the RNA or DNA. The information is primary, and the matter is secondary. But we cannot even get the matter right (requisite Carbohydrates, Nucleic Acids, Lipids, and Proteins), let alone the information. Where is the code? Where is the complexity of mediums to carry and translate the information to the growing cell? Put the DNA in your body end to end and you get a thread that reaches to the sun and back about one hundred times. This is not some abstract fact. It is the wonder of you - DNA.
>
> The best scientists in the world cannot create a single living cell even if supplied with all the chemicals in homochiral form

AND the informational code. They cannot assemble the cell. Never been done. But it may be done one day, "says Dr Tour.

Dr Stephen C Meyer elaborates on the relevance of mind and intelligence to the Origins of Life debate.

> During a session on the origin of life, a vigorous debate broke out about the implications of the information stored in DNA. All the scientists on the panel acknowledged that current theories of chemical evolution had failed to explain the origin of the genetic information necessary to produce the first life. Some of these scientists thought that origin-of-life research simply needed more time to devise an explanation within a standard materialistic framework. Others, however, thought that scientists needed to consider a radically new explanatory approach—one that recognized *the connection* between intelligence and the production of information.
>
> From The Return of the God Hypothesis

Summary:

According to the fossil record, life appeared as basic single-cell bacteria soon after water appeared in liquid form on planet Earth.

Primitive but incredibly complex.

We do not know how that happened, but we may one day.

But we do know why that happened.

With wisdom and incredible intelligence, God spoke!

CHAPTER 8

Genesis Day 4

The fourth day started about 1.8 billion years ago.
Genesis 1:14-19 CJB says:

God said, "Let there be lights in the dome of the sky to divide the day from the night; let them be for signs, seasons, days and years;

and let them be for lights in the dome of the sky to give light to the earth" and that is how it was.

God made the two great lights - the larger light to rule the day and the smaller light to rule the night - and the stars.

God put them in the dome of the sky to give light to the earth,

to rule over the day and over the night, and to divide the light from the darkness; and God saw that it was good.

So there was evening, and there was morning, a fourth day.

Planet Earth, the luckiest planet, is tuned for life from its beginnings. Doctor Gerald Schroeder, internationally renowned scientist and physicist, agrees. He explains the early Hebrew writings in Genesis:

> The earth brought forth life. Earth had within it the necessary properties for life to flourish.

And as I describe these events in Makarrata:

> Water arrives at just the right stage in the planet's development. Within a surprisingly short period, simple life forms begin, something modern science can still not explain. In the iron-rich, newly formed shallow oceans, Stromatolites, a mushroom-shaped aggregation of minute animals, emerge as one of the earliest life forms. They use the energy from the sun to produce oxygen through the process of photosynthesis. First, they extract the iron oxide, which is really rusty. Next, they take billions of tonnes of iron from the oceans, forming massive bodies of iron-rich rock. Some of this exists in vast areas of northern Australia to this day. This process makes the oceans suitable for other life. Then, these early life forms set about creating the oxygen-intense atmosphere we enjoy today.
>
> Our earth takes its first breath! Photosynthetic microbes produce an oxygen-rich atmosphere, which soon becomes transparent. As these biologically driven reactions proceed, the Sun, Moon, and stars, already visible from space, became visible on Earth as individual light sources. Genesis describes this event from an earthly viewpoint, which is made clear by referencing the Moon as a great luminary. The Earth is the only celestial body close enough to the Moon to see it.

Michael Chambers. *Makarrata. The Australians of Arnhem Land*

According to Dr Stephen Meyer, in his book, The Return of the God Hypothesis:

> In addition to fine-tuning the laws and constants of physics and the arrangement of matter and energy at the beginning of the

universe, physicists have discovered many other contingent, finely tuned features of the universe. For example, a life-permitting universe depends crucially on its precise expansion rate. Since the discovery of the red shift of the light coming from distant galaxies, astronomers have discovered that if the universe were initially expanding even a smidgeon faster or slower, either stable galaxies would not have formed in the universe because matter would have dissipated too quickly for galaxies to congeal or else the universe would have quickly collapsed in on itself. Cosmologists refer to the first scenario as the "heat death of the universe" and the second scenario as the "big crunch." Neither outcome is friendly to life.

Before Darwin, biologists attributed the beauty, integrated complexity, and adaptation of organisms to their environments to a powerful designing intelligence. Consequently, they also thought the study of life rendered the activity of a designing intelligence detectable in the natural world. Yet Darwin argued that this appearance of design could be more simply explained as the product of a purely undirected mechanism—namely, natural selection and random variation.

In 1953, when Watson and Crick elucidated the structure of the DNA molecule, they made a startling discovery. The structure of DNA allows it to store information in the form of a four-character digital code. Strings of precisely sequenced chemicals called nucleotide bases store and transmit the assembly instructions—the information—for building the crucial protein molecules and machines the cell needs to survive.

Like the precisely arranged zeros and ones in a computer program, the chemical bases in DNA convey instructions by virtue of their specific arrangement—in accord with an independent symbol convention known as the genetic code. Thus, biologist Richard Dawkins notes that "the machine code of the genes is uncannily computer-like." Similarly, Bill Gates observes that "DNA is like a computer program, but far, far more advanced than any software we've ever created." Biotechnologist Leroy Hood likewise describes the information in DNA as "digital code."

The information in DNA also defies explanation by reference to the laws of chemistry. Saying otherwise is like saying a newspaper headline might arise from the chemical attraction between ink and paper. Clearly, something more is at work.

Instead, we think intelligent design is detectable in living systems because we know from experience that systems possessing large amounts of such information invariably arise from intelligent causes. The information on a computer screen can be traced back to a user or programmer. The information in a newspaper ultimately came from a writer—from a mind. As the pioneering information theorist Henry Quastler observed, "creation of information is habitually associated with conscious activity."

Imagine you have a thought. It's in your head, your mind. You scribble that thought on a piece of paper. Now, it's not only a thought; it's information, but it is in a different medium. Next, you sit and enter your writing into a document on your computer. Another medium; more information. Now, you email it to a friend across the world. Another medium. Its downloaded, opened and understood. Same information, different medium. So, get this. The molecules in our body, the hosts of DNA, change every few years.

Yet, information *is retained and* passed on despite this ever-changing environment. Every molecule is continuing to change, yet the information is retained! We are dynamic structures, constantly renewing. What is the real me? It's not the matter, not your body; it's continually changing. That is the difference between brain and mind, as described by the Hebrew word, *Neshama*. Spirit!

Even extremely slight alterations in the values of many independent factors—such as the universe's expansion rate, the speed of light, the masses of quarks, and the precise strength of gravitational or electromagnetic attraction—would render life impossible.

Professor Sir Fred Hoyle argued, "A commonsense interpretation of the facts suggests that a super intellect has monkeyed with physics, as well as chemistry and biology." Many physicists now concur. They would argue that—in effect—these parameters appear finely tuned to make life possible because someone carefully fine-tuned them.

In addition, proponents of intelligent design insist that the case for intelligent design is based upon scientific evidence and established scientific reasoning methods—not religious belief or authority.

When I was a young boy growing up in New Zealand in the mid-20th century, I developed a lifelong love of music. All music; I embraced Rock 'n Roll, Country, Folk, Jazz, and popular music. But I was particularly fond of Classical music. Symphonies, concertos, sonatas, and opera. I love Beethoven, Mozart, Schubert, Rachmaninov, and all the greats. When I could, I would lock myself in the living room for hours and delve into my mother's grand collection of records, vinyl we call them today. The volume of her new stereophonic sound system was maxed out! My mother said I had perfect pitch, and though I think that was just mother, but I have an excellent ear, though never able to play any instrument or read a single note. Eventually, I would conduct privately, by ear, many of the great symphonies from memory. Woodwinds Strings and Percussion sections, I was enthralled! I knew every note for every instrument.

The symphony of life for every living cell is directed by a conductor (That we know to be DNA and RNA). A conductor uses a score –a written record of every note of music, in sequence, and by individual musicians, each with a different instrument. Complex, is it not? But do not be mistaken; the score (DNA) is not the source but *merely the proof* of an underlying mind or intelligence. DNA is not the source of life; it is simply a conductor following a script. It points brazenly and unashamedly toward an intelligent design.

The fine-tuning of the universe as a whole is better explained by an intelligent agent that transcends the universe, one that has the attributes that religious believers typically associate with God. The fine-tuning of the universe predates the beginning of time and the origin of the specified information that arises after the beginning of time, which is necessary to produce the first living organism.

In his book Science Returns to God, Gerald Schroeder explains:

> You see, if you win a lottery this week and then again next week, and then again the third week, chances are that before you collect your third winnings, you will be on your way to jail for having rigged the results. The probability of winning three in a row, or three in a lifetime, is so small as to be negligible. "But," you plead before the judge, "probability never says never. It was just a rare set of circumstances." It's true that probability never

says never, but all of physics, which means all of nature, is based on the understanding that the very, very, very unlikely never happens. Without this basic understanding, there is no foundation for any assumptions of physics or cosmology. With the universe we did not win just one lottery. We won at the choice of the strength of the electromagnetic force (which encourages atoms to join into molecules). We won at the strength of the strong nuclear force (which holds atomic nuclei together; were it a bit stronger then diproton and not hydrogen would be the major component of the universe, and no hydrogen means no shining stars).

Consider carbon, element number 6 in the periodic table. Before it comes hydrogen, helium, lithium, beryllium, and boron. After that comes nitrogen and oxygen, and the remaining ninety-two natural elements. Life as we know it is based on carbon. It is the only element that can form the long and complex chains necessary for the processes of life. Elsewhere in the universe life may be based on liquids other than water, but carbon is the necessary elemental jack-of-all-trades for life. It is also the essential Lego-like stepping stone for the production of the eighty-six natural elements heavier than carbon.

Summary:

God spoke to the planet, our Earth, and told it to "bring forth life," using all the elements necessary for life to flourish.

He did not specify a timeline, for God does not need one.

Life appears suddenly, mysteriously in the fossil record. Simple single-celled life yet incredibly complex, based on Information.

For millions of years, nothing much really happened.

Photosynthesis from the living organisms eventually purified the previously dense atmosphere, and the sun, moon, and stars became visible from Planet Earth.

But then, billions of years after the Big Bang, an amazing event occurred that would change life on Earth forever!

CHAPTER 9

Genesis Day 5

Day five started 750 million years ago.

The Bible says:

> God said, "Let the water swarm with swarms of living creatures, and let birds fly above the earth in the open dome of the sky."
>
> God created the great sea creatures and every living thing that creeps, so that the water swarmed with all kinds of them, and there was every kind of winged bird; and God saw that it was good.
>
> Then God blessed them, saying, "Be fruitful, multiply and fill the water of the seas, and let birds multiply on the earth."
>
> So there was evening, and there was morning, a fifth day.
>
> Genesis 1:20-23 CJB

The first animal life swarms abundantly in the oceans followed by large reptiles and winged animals.

The first multicellular animals, which have the basic body plans of all future animals, appear along with winged insects.

About half a billion years ago, with no apparent warning, an amazing event occurred on the luckiest Planet. According to the fossil record, seemingly from nowhere, and without earlier fossils, a whole range of unique new life appeared!

Gerald Schroeder explains:

> At the Cambrian explosion of animal life 530 million years ago, some 50 phyla (basic body plans) appeared suddenly in the fossil record. Only 30 to 34 survived. The rest perished. <u>Since then, no new phyla have evolved!</u> It is no wonder that the Scientific American Journal asked whether the mechanism of evolution has changed in a way that prohibits all other body phyla? It is not that the mechanism of evolution has changed. Our understanding of how evolution functions must change to fit the data presented by the fossil record directions. Macro-evolution, the evolution of one body plan into another—a worm or insect or mollusc evolving into a fish, for example—finds no support in the fossil record, lab, or Bible. Microevolution is the gradual evolution of a trait that only slightly alters the animal's morphology. It is observed regularly in farmyards and biology laboratories. It finds no dispute in the Bible.
>
> Could fifty genetically separate phyla have evolved these similar genes individually by chance? Twenty different amino acids are available to fill each of the 130 spaces on the gene. This means there are 20^{130} or 10^{170} possible combinations. There are one hundred million, billion, billion, billion, billion, billion, billion, billion, billion, billion, billion, billion, billion, billion, billion, billion, billion, billion, billion ways the amino acids can arrange themselves in those 130 slots on the gene. The number far exceeds the number of particles in the entire universe.
>
> Dr Gerald Schoeder. Evolution and Randomness

The evolution of a new species from an older one is <u>never</u> found in the fossil record.

Summary:

> *Soon after the arrival of water on the planet, life started abruptly, something science has failed to explain.*
>
> *There was no slow evolution – it just started, with complex cells and organisms soon afterward.*
>
> *Then 530 million years ago, it exploded.*
>
> *Since then, no new phyla have emerged.*
>
> *The only changes have been made through natural selection, which we will learn more about in the next chapter.*

CHAPTER 10

Genesis Day 6

Day six started about 250 million years ago.

In Genesis 1-24:31 the CJB says

God said, "Let the earth bring forth each kind of living creature - each kind of livestock, crawling animal and wild beast" and that is how it was.

God made each kind of wild beast, each kind of livestock and every kind of animal that crawls along the ground; and God saw that it was good.

Then God said, "Let us make humankind in our image, in the likeness of ourselves; and let them rule over the fish in the sea, the birds in the air, the animals, and over all the earth, and over every crawling creature that crawls on the earth."

So God created humankind in his own image; in the image of God he created him: male and female, He created them.

God blessed them: God said to them, "Be fruitful, multiply, fill the earth and subdue it. Rule over the fish in the sea, the birds in the air and every living creature that crawls on the earth."

Then God said, "Here! Throughout the whole earth I am giving you as food every seed-bearing plant and every tree with seed-bearing fruit.

And to every wild animal, bird in the air and creature crawling on the earth, in which there is a living soul, I am giving as food every kind of green plant." And that is how it was.

God saw everything that he had made, and indeed it was very good. So there was evening, and there was morning, a sixth day.

The Torah is clear: God's objective for life on Earth is mankind, and humans are made in His image, the highest form of life. It was not easy; many obstacles were in the way for millions of years. For instance, 250 million years ago, a mass extinction mysteriously occurred, decimating life and 90% of species disappeared from the fossil record. Why? We know God has intervened from time to time in Earthly affairs. Was life heading in the wrong direction? We will probably never know for sure.

This is what the [NASA Earth Science magazine](#) has to say about it:

> "It was almost the perfect crime. Some perpetrator – or perpetrators – committed murder on a scale unequalled in the history of the world. They left few clues to their identity, and they buried all the evidence under layers and layers of earth. The case has gone unsolved for 250 million years, that is. But now the pieces are starting to come together, thanks to a team of NASA-funded sleuths who have found the villain's " fingerprints " or at least one of the accomplices. The terrible event had been lost in the amnesia of time for eons. It was only recently that palaeontologists, like hikers stumbling upon an unmarked grave in the woods, noticed a startling pattern in the fossil record: Below a certain point in the accumulated layers of earth, the rock shows signs of an ancient world teeming with life. In more recent layers just above that point, signs of life all

but vanish. Somehow, most of the life on Earth perished in a brief moment of geologic time roughly 250 million years ago. Scientists call it the Permian-Triassic extinction or "the Great Dying" – not to be confused with the better-known Cretaceous-Tertiary extinction that signalled the end of the dinosaurs 65 million years ago. Whatever happened during the Permian-Triassic period was much worse: No class of life was spared from the devastation. Trees, plants, lizards, proto-mammals, insects, fish, molluscs, and microbes – all were nearly wiped out. Roughly 9 in 10 marine species and 7 in 10 land species vanished. Life on our planet almost came to an end.

NASA Editorial Team article. The Great Dying.

Life on earth had to start again, get back on track, so to speak. And with spectacular success, culminating about 65 million years ago with another mass extinction. The end of the dinosaurs and the start of the mammals. What happened? Probably a great asteroid from outer space. Why? Was life once more heading in the wrong direction? But for sure, without this particular event, we wouldn't be here – humankind would not exist.

In the following chapters, we will address the most common questions. How did life begin? Who was Adam? Is Evolution real, and did we evolve from the apes? Are dinosaurs mentioned in the Bible? But for now, about 6000 years ago, the centrepiece and culmination of all Creation appear suddenly and dramatically. Torah warrants the story to be so crucial that it explains the narrative twice, each from a different perspective. The reality of duality. Perhaps it happened like this?

The Earth, the luckiest planet, bows before The Living God, its Creator, and brings forth the Man and Woman, the culmination of the ages formed from the dust over millennia, and presents them to Adonai, the Lord.

"Behold, I have completed the task you set me. Here they are, male and female. I have bought them forth from the earth, as you directed!"

And God takes them and breathes into them new Life! A new creation. He shows them the wonders of the Planet they have inherited. That is now theirs to rule over!

And God saw that it was not just good, but very good!

At this moment, God breathes the breath of life (in Hebrew, the *Neshama*) that the created and the Creator become eternally linked. In this third and final act of Creation, God breathes the "soul of life" into the creature before him, and he becomes a human being. The first human beings are Adam and Eve. We are so intimately connected to the earth, and our roots are so entwined with that of the planet that the name chosen for the first human to represent mankind was *adamah*, Adam, meaning earth. Meaning dust. We should recall that every atom and particle of the planet that bought forth Adam had already been "recycled" by the universe from other galaxies, stars, and planets now long gone. Adam, meaning star dust!

In his book The Science of God: The Convergence of Scientific and Biblical Wisdom, Gerald L Schroeder explains.

> In each of the first two instances, Onkelos translates nefesh haiyah as a living animal, the literal meaning of the Hebrew.

For mankind, the Hebrew text has a slight variation, saying that the adam became "to a nefesh haiyah" (Gen. 2:7). Based on the addition of the word "to," Onkelos translates nefesh haiyah as "communicating spirit." And the Eternal God formed the adam dust from the ground and breathed into his nostrils the Neshama of life and the adam became a communicating spirit" (Gen. 2:7). The vital element that sets humans apart from all other animals is our immanent spirituality and our ability to share that spirituality with others. Archaeologists can never discover the fossil remains of Neshama. It is totally spiritual. That notwithstanding, archaeological evidence has confirmed our biblical heritage. As it is written: "The truth shall spring from the Earth" (Psalm. 85:12)

What an astounding revelation; we who are made of stardust are communicating spirits!

Once the *Neshama,* the soul of human life, appears on the earth, the Torah changes its perspective, from looking forward in time from Creation to looking backward with a human viewpoint from the Planet Earth. Suddenly, a new clock starts ticking, one unique to humans on this planet, that measures the passing of time. The beginning of recorded time. The Jewish calendar calculates the year by adding up the generations since Adam began to walk with God. Because time was measured differently, the six days of creation are counted separately. As we measure the time from Jesus, the Jews continue as they have done for nearly six thousand years to calculate from Adam, the first human. Currently, the Jewish year is 5785.

It is interesting to look back on history and realise humanity's incredible leaps in the last few thousand years. As the first human beings, communicating spirits, *Neshama* made its presence known. In The Science of God, Gerald Schroeder explains:

> The stimulus for the emergence and development of writing was the need to record economic transactions. The start of writing relates to the need for bartering, which relates to the formation of large cities. The question remains: why did large cities form at that time? With the invention of agriculture, the land could already support large populations. It is, however, unlikely that a population explosion was the sole cause of the start of large cities. Hominids invented agriculture almost four thousand years before the creation of Adam. For those thousands of years, they did not live in large cities. A key element facilitating the transition from village to city was missing. I propose that the missing factor was the *Neshama*.

Gerald Schroeder continues.

> The difference between animals and humans is further underscored by the fact that God assigned to humans the responsibility of stewardship "over" the animals (Genesis 1:26). Finally, that difference is also the focus in Genesis 2:18–24, where the way in which the narrative is structured shows that the naming of the animals is to be read in the context of finding a helper for Adam. The lesson is that no helper was found fit for (or corresponding to) Adam among the many animal species then existing – therefore, including, be it noted, whatever nonhuman hominids may or may not have existed at the time. It is more than interesting that, according to the Bible, the first lesson Adam was taught was that he was fundamentally different from all other creatures.

Summary:

The Earth bought forth life, plants, creatures, and animals of every kind.

One of those species was hominids, homo sapiens.

God chose Adam and Eve from them and created a new Spirit in them, the neshama, a New creation. Spiritual beings created in the image of God.

He blessed them and separated them from all other living things. He gave them dominion over the whole planet and every living thing.

And He saw it was very good.

CHAPTER 11

What about Darwin?

Besides the scriptures, Darwin's great work "The Origin of Species by Natural Selection" might well be the most significant literary work in the annals of recorded history. Recently, it has rivalled the Bible in its popularity of ideas and influence!

Darwin proposed that all life, from the simplest to the most complex, originated by chance and evolved from a common ancestor or ancestors employing natural selection and the survival of the fittest.

Sir Julian Huxley, an eminent British supporter of Darwinian evolution, declared that the evolutionary dogma it spawned was the most potent and comprehensive idea ever on earth. (Emphasis on the was!)

The 20th century cannot be understood apart from Darwin's view of evolution.

It intruded on and influenced almost every field of human and scientific endeavour. Its greatest challenge remains the catastrophic effect it has had on Judeo Christianity. It was the great myth of the last 150 years.

"If evolution is reflective of the laws of science, then Genesis must be reflective of the flaws of scripture!" So say the detractors of the Bible.

But those days are now long gone. Well into the 21st century, the theory of evolution (and it was only ever a theory) is starting to unravel.

If you will live your life thinking you are the unthinking chemical product of some primordial soup, think again. You have a Creator; this fact alone should change how you view and live life!

"Zombie Science" is materialistic science masquerading as empirical science. It is false, a copy, a delusion.

Evolution can have many meanings. It can mean change over time. All things change over time. That's called entropy or corruption. More about that later. Another meaning of evolution can mean there is a history to life on earth. That there are things that aren't with us anymore. Some things, like human beings, are here but weren't always here. In those senses, evolution is not controversial. Charles Darwin's explanation for all those changes was purely materialistic. He denied there was any design. He said it was all chance. His theory won out not because of evidence but because it fit the materialistic temper of the time. Micro evolution plus time equals macro evolution. That hypothesis is starving for evidence. It struggled in Darwin's time, and still today.

The primary and most important meaning of science declares that through empirical investigation, we formulate a theory or hypothesis and test it against the evidence. If it fits the evidence, we keep it. If it doesn't, we throw it away. Or modify it. That's empirical science, and that's how science should be done. For empirical science, the highest value is truthfulness, but for materialistic science, the highest value is survival of the fittest. Of course, there have been scientists in the past who put self-interest above the search for truth. However, many scientists still consider truthfulness to be the highest value.

Author and scientist Johnathon Wells explains.

There are other meanings of the word science, like "consensus science." Which is the majority opinion of current practicing scientists. Which often turns out to be wrong if you look back at the history of science.

Look at what happened to science under Hitler and the Jewish Holocaust. That was a result of "consensus science" that declared Jews to be on the lowest rung of humanity along with Australian Aborigines and black Africans. Barely human, according to Hitler! The actual route of racism in society comes through Darwinian evolutionary theory. I often remember the beautiful black family in Capetown when I was a boy. That sort of apartheid thinking stems directly from Darwin. In How to Defeat Zombie Science on Discovery Channel, Johnathon Wells says:

> The third and more disturbing definition is the search for natural (materialistic) explanations for everything. Now it's okay to search for natural explanations, but if you insist that all science must be based on materialistic fact, then that's basically a philosophy and not empirical science. Zombie science is telling materialistic stories even though they are dead! Those ideas are walking dead! They don't fit the evidence. They are empirically dead but they keep stalking the halls of science and education.

When I was young, I was bombarded with a diagram with a series of images showing an ape-like creature walking stooped with hanging arms that gradually morphed into a human being several stages later, the result of many "missing" links! It is now roundly repudiated as false by scientists worldwide but continues to be used in textbooks and lectures. According to Johnathon Wells, this is just one of many examples of zombie science. This idea lives long after its death, deceiving and affecting modern thinking and education. His recent book gives many examples of zombie science. Here is a list of discredited and debunked theories.

- Darwin's Tree of Life: A branching tree diagram used to illustrate the notion of descent with modification of all living things from common ancestors; False.
- Homology in Vertebrate Limbs: Similarities in limb bones are used as evidence that vertebrates (animals with backbones) are all descended from a common ancestor; False.
- Haeckel's Embryos: Drawings of similarities in early embryos used as evidence that all vertebrates (including humans) evolved from fish-like animals; False.
- Archaeopteryx: A fossil bird with teeth in its mouth and claws on its wings, often cited as the missing link between ancient reptiles and modern birds; False.
- Peppered Moths: Photos of moths resting on tree trunks, supposedly providing evidence for evolution by natural selection; False.
- Darwin's Finches: Thirteen species of finches on the Galápagos Islands that are used as evidence for the origin of species by natural selection; False.

- Four-Winged Fruit Flies: Fruit flies with an extra pair of wings that supposedly provide evidence that DNA mutations provide the raw materials for macroevolution; False.
- Fossil Horses: Fossils once used to show that evolution proceeds in a straight line and later used to show that it doesn't. False.

All these "theories" have been debunked by modern empirical science. Science is based on facts, not fairytale theories.

With the fantastic advances in biology, archaeology, astrophysics, and many other branches of science, the evidence against Darwinism is slowly mounting to the point where many ardent evolutionists no longer subscribe to the theory.

In Zombie Science. More Icons of Evolution. Johnathon Wells says:

> When I was a boy growing up in northern New Jersey, a lake near our house would freeze hard in the winter, and I would skate on it with my friends. As the weather grew warmer in the early spring, the ice would become honeycombed with pockets of meltwater. Although the spring ice still looked thick and solid, my friends and I knew it was no longer strong enough to hold our weight, and we stopped skating on it. Today, evolutionary theory is like spring ice. It still covers the lake, and it still looks solid to many people. But it's honeycombed with melt-water. It can no longer carry the weight it once did. Summer is on the way.

Darwin's discredited theory of evolution is skating on thin ice. A new day is dawning, where the icons of evolution past give way to empirical science and the weight of evidence. A new day is dawning. As The Beatles sang nearly sixty years ago:

"Here comes the sun... Little darling,
I feel that ice is slowly melting.
Little darling, it seems like years since it's been clear.
Here comes the sun, here comes the sun,
And I say it's alright.

Summary:

In the last century, Darwin's theory of evolution reigned supreme in scientific and religious circles.

Recent advances in science and biology refute the bedrock of the theory that natural selection leads to new species. There is no proof of that in the fossil or natural world.

In the next chapter, we will discover the actual function of natural selection.

CHAPTER 12

Mico Evolution and Nanotechnology

So far in this book, we have looked at the macro, the big picture, the universe, and the formation of galaxies and planets like our own. But there is another world to explore, equally as challenging and exciting. The quantum world of microtechnology is the study of life at its smallest level.

According to The European Commission:

> Nanotechnology refers to the branch of science and engineering devoted to designing, producing, and using structures, devices, and systems by manipulating atoms and molecules at nanoscale, i.e. having one or more dimensions of the order of 100 nanometres (one ten thousandth of a millimetre) or less. The applications of nanotechnology can be very beneficial and have the potential to make a significant impact on society.... Nanotechnology has already been embraced by industrial sectors, such as the information and communications sectors, but is also used in food technology, energy technology, as well as in some medical products and medicines. Nanomaterials may also offer new opportunities for the reduction of environmental pollution.

Dr James Tour is an internationally acclaimed scientist and one of the world's leading exponents of nanotechnology. He was instrumental in the discovery and commercial application of graphene. This robust and versatile material is less than one atom thick! Graphene is one of the most valuable materials and has numerous commercial innovations.

He and his team's contributions to modern science include

- Treatment of Traumatic Brain Injury, stroke and dementia with carbon Nanoparticles
- Healing of completely severed spinal cords with graphene nanoribbons
- "Tattoo Therapy" with carbon nanoparticles to alleviate autoimmune disease
- Drug delivery with carbon nanoparticles
- The treatment of anti-biotic resistant bugs.

He and his team are also pioneers in the field of building nanomachines. These incredibly complex, microscopic machines operate inside the cell. They are so small you can fit 50,000 of them across a human hair! (How is that even possible?)

Dr James Tour also teaches in the field of the Origins of Life. I highly recommend viewing a presentation he made recently on the Discovery Channel.

He reminds us that unlike other fields of science, which have made enormous advances in recent years, in the Origins of Life field, nothing has changed since 1952. According to Scientific American magazine,

> … in 1952, Stanley Miller and Harold Urey simulated the conditions of early Earth by sealing water, methane, ammonia, and hydrogen in a glass flask. Then, they applied electrical

sparks to the mixture. Miraculously, amino acids came into existence amid the roiling mixture. It was a big deal.

The experiment created a furore among the origins of life scientific community. Still, like every subsequent attempt to prove the origin of life on earth, according to Dr James Tour, they have all failed. Yet, like the theory of evolution, they continue to make claims that are not replicable by empirical science.

According to Zombie Science. More Icons of Evolution by Johnathon Wells

> In the 1990s, however, a form of opposition to evolution with no position on the age of the Earth rose to prominence in America: Intelligent Design (ID). According to ID, it is possible to infer from evidence that some features of the natural world, including some features of living things, are better explained by an intelligent cause than by unguided natural processes. So, unlike young Earth creationism, ID is based strictly on scientific evidence rather than a combination of scientific evidence and Bible-based arguments. But it has caused a furore among many followers of Darwinian Evolution.

Why this hysteria over ID? Because if the evidence shows that even one feature of living things is due to intelligent design instead of unguided natural processes, the whole edifice of zombie science comes crashing down.

In his 2009 book Signature in the Cell, philosopher of science Stephen Meyer argues that the complex information in biological molecules cannot result from unguided natural processes such as the spontaneous aggregation of chemicals. The only known source of large amounts of complex information is intelligence. Therefore, Meyer concludes the origin of life requires intelligent design.

Intelligent Design is based on a theory that is growing in stature and popularity, is purely scientific, and relies on empirical science. It is not religious, but the Bible has little dispute with its findings. It points toward God as the Creator.

Summary:

> *The old theories and assumptions of evolution that were not based on empirical evidence are being swept away by a new wave of scientific knowledge and understanding.*

> *There is a growing awareness that Darwin's theory of progressive evolution is skating on thin ice.*

> *More and more scientists are coming to realise that Intelligent Design is necessary to better understand our natural development.*

CHAPTER 13

Devolution and Secrets of the Cell

For thousands of years, we, as creative human beings, have designed and made machines to help us enjoy a better, more productive life. We are a machine culture, from the simple stone age animal trap or boomerang to the modern autonomous robot. Nowadays, some manmade machines are staggering in both size and complexity.

Professor Michael Behe serves as a Professor of Biochemistry at Lehigh University in Bethlehem, Pennsylvania, and as a senior fellow of the Discovery Institute's Centre for Science and Culture. He is the author of several bestselling books, Darwin Devolves, The Edge of Evolution, and Darwin's Black Box. Most of the following ideas can be found in more detail in his excellent documentary series Secrets of the Cell.

Michael Behe explains:

> Microbes are single-celled organisms. Advanced forms of life have trillions of cells. Inside a cell is a bustling operation of interconnected machinery so sophisticated that it would shame any high-tech factory. This intricate machinery gives the cell remarkable abilities to process energy, execute generic instructions, and replicate, all with unbelievable efficiency.
>
> Question? Exactly how did the cell get to be so complex?

We all love cars for their beauty and performance. But only mechanics understand what's inside. They have up to 30,000 working parts. In a way, a cell biologist is a lot like a mechanic. Each cell has an astonishing number of intricate parts, way more than a car. We study what the parts are, how they work together, and even which parts of the cell can be left out without causing too much harm. If you understand the workings of the cell, you can understand a lot about all living things. What makes them thrive, what makes them fail and what causes them to evolve

The Secrets of the Cell. Michael Behe. Discovery Science Institute

Let's look at the problem of complexity. Our bodies are made from about 40 trillion individual cells. But actually, we could reduce the number of parts and still perform quite well. We could lose a hand or an ear, or even a kidney, and still work okay. What if we look at an electric drill. Some of the functions we could do without, like the battery gauge or the LED light, but it would still work okay. But how far could we reduce the complexity and still have a working drill? Let's look at another example. A simple wooden mouse trap.

In Secrets of the Cell. Discovery Science Institute, Michael Behe explains

> The complexity of this machine entails five parts; the platform, the spring, the hammer, the holding bar, and the catch. Now, if I reduce the number of parts from 5 to 4, if I take away the holding bar or the spring or any other part it wouldn't work. Every part is essential. I call this <u>irreducible complexity</u>. So, does this apply to biology, too? Do we have examples of

irreducible complexity in life? Consider a bacterium, a single-cell microbe. Some bacteria have a flagellum, which is like an outboard motor on a boat. It rotates to propel bacteria through a liquid, and it's necessary for their survival in order to swim to search for food.

The flagellum has a number of parts. A drive shaft, a universal joint, a rotor, bushings and stator. Even a propellor, a clutch and braking system. The motor of the flagellum has been clocked at a hundred thousand revolutions per minute, and as with the mousetrap, removing even one of the components of this elegant machine destroys its function and renders the bacterium shall we say, dead in the water.

The flagellum is irreducibly complex. Now we know how an outboard motor is assembled in a factory, but how is the bacteria flagellum developed in nature. Keep in mind it is irreducibly complex; all the parts have to be in place, or it does not work. Could it have been developed blindly in stages? Let's say in the distant past, the bacteria first gained a drive shaft, later on, a rotor, then a stator, and then the rest, and then the parts all got in sync like a high-performance engine.

Darwin's theory of evolution says that with enough time, any living organism, no matter how complex, will slowly develop into something more complex by random, unguided forces. Or does it?

Let's look at the animal kingdom. Insects, for instance. They are actually mechanical marvels. Studying them inspires modern science in the ever-expanding field of biomedics, which is making advances by mimicking biology.

In Secrets of the Cell. Michael Behe explains:

> For example, studying the eyes of flies led to the development of advanced optical lenses. The physics of dragonfly wings allowed increased efficiency of wind turbines. Cockroach mechanics paved the way for increased agility in robotics. A honeybee algorithm helped optimize the internet and the tiny plant hopper (known as *Fulgoroidea*) is helping scientists to maximize the strength and rotation of gears. What gears? I thought humans invented gears. Not so. It turns out that biology is far ahead of our engineering. The plant hopper launches like a rocket and can jump more than a hundred times its body length. That's equivalent to a human jumping across not one but two football fields! Using special cameras scientists at Cambridge University discovered that the plant hopper coordinates the motion of its legs by means of incredibly precise gears! Making long leaps accurately requires that the legs be in perfect sync. If one leg flexed even a fraction of a second before the other the insect would lose power and tumble erratically.

For the last 150 years, biologists have thought they have figured out Darwin's theory of evolution. That life evolved into higher species through simple variation and determined which would survive. Then, over time, we change from one species to another. That is only partially true.

Here is just one example; the arctic polar bear. According to Michael Behe, here is how it may have evolved.

> Many thousands of years ago, a bear cub was born with a mutation in its DNA. The change gave the lucky bear the ability to eat a high-fat diet like seal blubber. So, the mutant

bear and its descendants could hunt seals for food. Generations later a baby bear was born with another mutation that altered the colour of its hair from brown to white. It was useful in the snow as camouflage, allowing it to sneak up on prey. So, variation in the ancestral species of the brown bear followed by natural selection for the harsh arctic surroundings were responsible for building the magnificent polar bear in small steps. That is natural selection. We can see fascinating evolutionary changes in many species, but modern science knows something that Darwin didn't know. That changes that we can see in species are driven by molecular changes in genes and DNA. So here is the big question. What exactly are those helpful mutations in DNA? How exactly do genes change? If brown bears can give rise to polar bears does that mean we will see future genetic changes making new species? Will we humans mutate into super-powered X-men?

Secrets of the Cell. Michael Behe. Discovery Science Institute

Dogs are one of the most diverse species on the planet. Geneticists claim they all have the wolf as a common ancestor. Modern dogs are not a new species; they are a new breed. They can appear completely different from each other, but they are still dogs, and they share much of their DNA with the original wolf. How does this work? Let's take a look inside the cell. In the nucleus we find DNA. This consists of a long double chain of chemicals twisted like a helix. DNA is found in all living cells. These chemicals are named A, G, C, and T. The DNA sequence is a combination of each of those letters. There are trillions of available combinations. Its information. Its speech. It's the longest word in the world. That's why many leading scientists, starting with Sir Fred Hoyle, claim that there has just not been enough time since life on earth began for life as we know it to evolve through mutations and natural selection. Mathematicians like Professor John Lennox have calculated that it would take trillions of years for such a process to occur. Not the few million years that have elapsed since life first started.

All DNA has these four-letter characteristics. Some sections are called genes. They contain the instructions for making the proteins that form the cell. But *do not claim*, as some do, that intelligence and, therefore, mind resides here. According to Professor Dennis Noble, an acclaimed biologist, DNA, while wonderous, is the messenger, not the author.

Genes give the dog its traits. Michael Behe continues:

> We have seen how mutations in DNA have mutated to form a polar bear. But it's still a bear. Likewise with dogs. A mutation in a gene occurs when there's a change in its sequence. Often this results in breaking one of the instructions forming the animal and its behaviour. In this case the mutation was in the specific gene that determines the hair colour. The disabled gene is unable to make coloured fur so the mutant fur is white.

> The change in polar bears happened long ago. A similar process happens in us humans to determine the colour of our skin.

> Wouldn't it be great if we could observe mutations today? Well, we're in luck. Critical work has been done in the laboratory of Richard Lenski, who works with a species of bacteria called E coli. His lab has been growing generations of E coli in laboratory flasks for 30 years and has witnessed millions of generations of E coli. In the early 1990's, they watched the

descendant bacteria begin to grow faster than their ancestors. That was great news for the happy E coli. But why was this happening?

A decade later, they determined the mutation that caused it. They found that a specific gene in the E coli had been destroyed. Losing that gene ended up helping the bugs to grow faster. Then they looked at a dozen other helpful genes in the E coli and saw that they, too, had been broken. Mutations had left the genes either crippled or completely disabled. But that seems odd. How can breaking a gene help the organism?

Secrets of the Cell. Michael Behe. Discovery Science Institute

Imagine you are out in your car and have to reach your destination on just one tank of petrol – or you'll die! Given the weight of your vehicle, you'll never reach your destination on just one tank. So, you need to shed excess weight. You toss out the bonnet, the passenger seat, floor mats, roof rack, the spare tyre, and even the cigarette lighter. The lighter, degraded car gets you to your destination, and you survive. But the car is not the machine it once was. It is much diminished.

So, by breaking genes, the resulting mutations helped E coli grow faster in the lab environment but as a degraded organism. But what about dogs? Just recently, scientists have realised just how many mutations affect the character and behaviour of different breeds of dogs. These are <u>not</u> new genes. They have not evolved. These genes are broken or damaged. They are degraded. Some dogs are more muscular; that comes from a broken gene. These broken genes exhibit familiar characteristics. Short tails. Friendly dispositions. Many shades and colours.

When we were young, my wife and I both trained sheepdogs. There was some serious competition between us, which still lingers! Both our dogs were a cross between two breeds; Border Collies from Scotland and Kelpies from the Australian outback. Her dog was called Gunna for all the things he was going to do and came from a long line of Kelpies bred by her father and generations prior. It was widely known that there were dingos in the bloodline. Many different characteristics from many other breeds. All the result of defective genes.

But that dog will never return to being a dingo; any more than a polar bear will turn brown and return to eating berries! The flying fish of my youth will never become a regular fish again. Those days are gone, and those genes are broken, never to be repaired. Like that car, the bits required to make it whole are left far behind.

In the documentary Secrets of the Cell, Michael Behe explains:

> In short, helpful mutations are not an upgrade. Getting a new smartphone is an upgrade. It has completely new features and is faster. Mutations don't install new features in the DNA. They only make changes to existing ones. A mutation is like disabling the GPS on your phone. It might save the battery but doesn't add a new function.
>
> We're told that random mutations are the main driver for evolutionary change and that evolution is responsible for upgrading lower forms of life to higher forms of life. Yet the latest scientific results show new species are made by breaking genes, by <u>devolution. That is the opposite of evolution!</u>
>
> Is there some unknown x-factor that boosts the capacity of evolution to gradually generate higher and higher life forms?

That accounts for the complexity of life? What was a mystery to Charles Darwin is now well understood. We know that the cell encodes and transmits information that regulates the size and shape of living things. And we know that the cell is comprised of insanely complex machinery. What's more, much of it is irreducibly complex. Like the mousetrap, all the parts are necessary for an irreducibly complex system to work.

Irreducibility is evidence that <u>all the parts were formed intentionally</u> with the end purpose in mind, not randomly over time.

Darwin's explanation for complexity was evolution, where random minor variations over many generations progressively led to life of greater and greater sophistication. However, science recently discovered that variation comes from genetic mutation and that helpful mutations usually break genes. That's not evolution, that's devolution.

So, what is the evidence for some x-factor that accounts for biological complexity? Let's think about it.

Let's say you are wandering through one of Australia's magnificent paperbark forests. Surrounded by life, both great and small. Your footsteps are cushioned by the dense carpet of fallen leaves, twigs, branches, and logs. You notice some of them are hollow, a natural process usually involving termites. But then you come across something weird and entirely unexpected. A long, straight piece of timber, and like some of the logs, completely hollow. You pick it up. It's a wooden tube, about as tall as you. Even more amazing – it has patterns carved and painted on it. It's a didgeridoo, or as it's called in Arnhemland, a yidaki. In the hands of a skilled musician, it can make wonderful music. So, what could you conclude from examining it? There is no apparent evidence to show when or who made it. Or how it was made. Yet, it is different from its surroundings. It speaks to us of planning and design. So, we do have evidence that it was created by an intelligent being. A mind was at work.

Summary:

As we look at the natural world through the perspective of both the Bible and Science, it becomes increasingly apparent that a mind is at work. An intelligent being. A designer.

Just a simple reading of the Bible will identify that Mind, that Intelligent Designer. That Creator.

CHAPTER 14

Where Did Adam Come From?

The appearance of humans in the 3500-year-old biblical story is best understood through the eyes of the ancient Jewish scholars who were not influenced by modern science or civilisation. Nachmanides (1194-1270), known as the Ramban. And Maimonides (1138-1204.)

Gerald Schroeder says,

> The biblical word, creation, implies a partial withdrawal of God's overt influence. In Hebrew, this concept is *tsimtsum.* That the creation described in (Gen 1:1) implies that God withdrew part of God's undifferentiated unity and allowed physical complexity to appear: time, space matter, the laws of nature. The creation of animals (Gen 1:21) relates to the creation of the soul of animals, (the *nefesh*) in Hebrew, and gives animals the ability to choose, to learn how to manipulate a maze. The creation of Adam (Gen 1:27) grants a further divine pull back, allowing us free will, the soul of humanity, the *Neshama* in Hebrew.
>
> Gerald L Schroeder. Evolution Bible Style.

It is clear that in the Genesis account of Adam and Eve, there are references to other "beings."

> And with that knowledge they describe the old/young age of our universe. They talk about "beings" that we today would refer to

as hominids, beings identical to humans in shape and in intelligence, lacking only the soul of humanity, the *Neshama,* to make them human. According to these ancient biblical commentators they walked the earth at the time of biblical Adam and before. "Cavemen" were never a theological problem to these ancient commentators.

Gerald L Schroeder. God According to God.

Many scientific papers in the technical literature have admitted that there is a severe gap between human-like fossils in the fossil record and our supposed ape-like ancestors.

The evidence? An eminent Harvard palaeontologist, Stephen J Gould, said, "...most hominid fossils, even though they serve as a basis for endless speculation and storytelling, are fragments of jaws and scraps of skulls."

The record is sparse and fragmented, with regular spurious claims of "the missing link" appearing regularly in the world media, only to gradually fade away as the actual facts are known. I'm sure you will remember many of these claims. The following is from The Unique Origins of Fossils in the Human Record. Casey Luskin, PhD. Discovery Channel.

> The scientific literature reports an explosion or rapid increase with punctuated change with approximately double the size of the brain at the first appearance of homo around two million years ago. Significantly and dramatically different. It was like the Big Bang for humans.... a major 2015 review of hominin evolution said that the evolutionary sequence for most hominin

lineages is unknown. Most hominin taxa, particularly early hominins, have no obvious ancestors.

There is nothing in the fossil record to indicate a previous ancestor. And indeed, no link to the ape family.

> The human race has unique and unapparelled moral, intellectual, and creative abilities, and there are many more obvious features of human beings that are exceptional. Our complex language, the complex tools that we make, use of fire, clothing, our ability to do abstract reasoning, are all traits that only humans have.

And so they were transformed when the Creator took two of these hominins and formed a new creation, the *Neshama*.

Gerald Schroeder explains,

> Humans have a source of pleasure not evident in other animals. It arises from the *Neshama*, our link to an all-encompassing unity that underlies what superficially appears to be a diverse and multifaceted universe. The *Neshama* whispers to us of a pleasure that transcends our limited physical existence. The decision-making program of the human *nefesh* now has two sources of information to consider as it strives for pleasure: the desires and needs of the body and the spiritual goals of the *Neshama*. How I choose to achieve my pleasure determines the quality of my person.
>
> Gerald L Schroeder. Evolution Bible Style.

To engage in mathematics, music, poetry, religion, and reasoning, I would add an amazing capacity for fiction. To tell stories. To create myths and legends while just sitting around a campfire.

Myths that foster friendship, fellowship, and cooperation among humans do not result from pure chance evolution …. but reflect the state of human society as if it was intended by a benevolent designer. As Casey Luskin PhD explains

In fact, there are many aspects of human cognitive and language abilities that go beyond the requirements of a species just to survive and reproduce. (Think) on the African savannah two to three million years ago, then (think) of Michelangelo painting the Sistine Chapel or Oscar Schindler who sat risking his own life to save people outside his own tribe; the Jews! Or Albert Einstein who was able to ponder the deep secrets of the universe. None of these traits would help you in any way survive and reproduce on the African savannah.

These are not the skills or traits of the animal kingdom. These point to a higher calling; a greater destiny, shrouded in the glory and intentions of a Creator. The first task God set Adam to perform was to name the animals in his kingdom. In other words, he was a higher life form than the rest. He was different. He was a communicating spirit!

Summary:

> *There is no evidence in the fossil record to suggest a link between the ape family and homo sapiens.*
>
> *Adam and Eve were chosen by God to become the New Creation. They were the first human beings about 5500 years ago.*

They were set apart, above the animal kingdom, infused with the neshama. They became communicating spirits. They had a personal relationship with God.

They were destined for eternal life.

So are you!

CHAPTER 15.

What About the Dinosaurs?

Why doesn't the Bible talk about dinosaurs? Well, it does! It's a question posed by believers and sceptics alike. The former is to better understand God's role in the world; the latter is to challenge the Bible's authority.

In his book, The Science of God, Gerald Schroeder addresses this question in an appendix. "Well, What About Dinosaurs?" He tells a personal story from his own experience. Here is an abridged version of his explanation.

> Why doesn't the Bible mention dinosaurs? Of course, the Bible has no obligation to mention them. It doesn't mention bananas or oranges, either. No one seems to have a problem with the other omissions, just with dinosaurs. There's a logic to the question. Dinosaurs represent a dramatic, tangible, even newsworthy fact by which science seems to challenge the Bible as an ultimate source of truth. Fortunately, the Bible does mention dinosaurs, though not by that name. Dinosaur is a fairly modern word, a composite of two Greek words meaning terrible lizard. That is a close approximation of the Bible's description of these beasts. Though I'd been asked about dinosaurs many times, I still never expected to hear the question in China, and certainly not three miles into the lush farmland of Jiangsu province, a locale that serves as the region's bread and tea basket.

Schroeder, Gerald L. The Science of God: The Convergence of Scientific and Biblical Wisdom.

Using his background in nuclear physics, Gerald and a friend travelled to a remote Chinese province to significantly increase the farmed fish population in the tropical and semi-tropical ponds that dot the landscape in China and Southeast Asia. A project that was spectacularly successful that bears fruit to this day. (Or should that be fish?)

He continues:

> It may have been the rush of emotion aroused by that view, coupled with the guaranteed privacy of the location that prompted my Chinese colleague to risk a question about religion. He certainly had never even hinted at the subject before. Chinese communism is so purely a materialistic culture that religion seemed to have no place. For all I knew it may have been illegal as well. We had stopped to catch our breath and admire the view. Without facing me he said "My grandmother has become a churchgoer. She never was before. She told me she believes the Bible is true.

Schroeder, Gerald L. The Science of God: The Convergence of Scientific and Biblical Wisdom.

Especially in the agricultural areas of China, Christianity is growing exponentially, and he was not surprised at the question. However, because of the strict regime imposed by the ruling Communist party, he felt that by answering, he could put his freedom at stake. He was not totally surprised by the question – it was common in the

Western world, and recently, there had been numerous finds of dinosaur remains in the Gobi Desert in China's north.

Gerald tells the story.

> It caught Zeng Li by surprise when I showed him where dinosaurs are mentioned in the Bible. In Genesis 1:21, we are told that on day five, God created the basis for all animal life. Among the categories of animals listed is one named *taninim gedolim*. *Gedolim* means big, and so we read "the big *taninim*.
>
> Pick up five different English translations of the Hebrew Bible, and you're likely to find five different meanings for the word *taninim*: whales, alligators, sea monsters, even dragons. Yet *taneen*, the singular of *taninim* is a word that appears elsewhere in the Bible, and its meaning is known.
>
> In Exodus 3, the Eternal spoke to Moses from the burning bush and told him to return to Egypt to lead the enslaved Hebrews to freedom. Moses felt incapable of the task and so the Eternal gave him several signs, one related to his shepherd's staff. When Moses was told to throw his staff on the ground, "it became a *nahash*" (Ex. 4:3). *Nahash* is the Hebrew word for snake. After Moses' return to Egypt, when Pharaoh asked for a sign, Moses' staff was again thrown to the ground and "became a *taneen*" (Ex. 7:10). Why didn't it become a nahash, a snake?
>
> And just five verses later, the Eternal tells Moses: "Get to Pharaoh in the morning, behold he goes to the water, and stand by the river's edge and the staff which turned into a *nahash* take in your hand" (Ex. 7:15). It's the same staff. The change is first referred to as a *nahash*, then as a *taneen*, then as a *nahash*.

We know that *nahash* means snake from its use elsewhere. *Taneen* must be a general category of animals since it appears in the creation chapter of Genesis where, other than Adam, only general categories of life are listed. So, *taneen* must be the general category within which *nahash*—snake—falls.

The general category for snakes is reptile. Thus, Genesis 1:21 translates as: "And God created the big reptiles…" The biggest reptiles were the dinosaurs. But the author of Genesis did not specify dinosaurs directly, because that would have been inconsistent with the pattern of the chapter.

The entire account of Genesis is stated in terms of objects known or knowable to the myriads of witnesses present at Sinai 3,300 years ago. Dinosaurs were not part of their world. But the hint and so many other hints were there in the text for later generations to discover.

"That discussion about dinosaurs was the first of many I had with Zeng Li. Over the years, in the privacy of that ridge before the ponds, we spent hours resolving supposed conflicts between science and the Bible. Years have passed since my last visit, and Zeng Li now joins his grandmother at religious services when research time permits. But for many persons the debate that underlies the question of dinosaurs remains."

Schroeder, Gerald L. The Science of God: The Convergence of Scientific and Biblical Wisdom.

Extensive beds of fossils record the rise and demise of these amazing animals. There were swimming dinosaurs, running dinosaurs, and even a form of flying dinosaurs. At about the same time, mammals appear in the fossil record. 65 million years ago, dinosaurs and mammals co-existed, but on anything other than equal footing.

Dinosaurs ruled the roost, getting bigger and more brutal, reaching sizes to rival today's great blue whale, while mammals failed to grow any larger than a few kilograms. Then it all changed. We think a meteor punched through the atmosphere, striking the earth with great force. There is a gigantic crater in the Caribbean, hundreds of kilometres across, that scientists believe resulted from that near-fatal blow. Dust and debris shrouded the planet. The sun was blocked for half a year, and there was a great darkness over the earth. Temperatures dropped to freezing. The dinosaurs all but died out. The saltwater crocodile remains one of the few to survive. Some of the smallest mammals survived. And life on earth began again.

Gerald Schroeder continues:

> In science, we search for the "how" of the universe. We study the world and hope to learn the laws by which it functions. But at the end of the day, the questions we all ask relate not to the how but to the why of existence. Is there meaning to our lives that transcends the splendour we see about us? Something that goes beyond the physical? And if so, how can we probe that meaning? So much of life seems to hang upon uncontrollable events. Was it chance or teleology (the explanation of phenomena in terms of the purpose they serve rather than of the cause by which they arise) that killed the dinosaurs?

And what of the meteor that almost killed us all?

Schroeder, Gerald L. The Science of God: The Convergence of Scientific and Biblical Wisdom.

The Luckiest Planet

In March 1989, The Luckiest Planet endured another close call. There were no alarms, no panic, and no one saw it coming. A smaller giant meteor than the one that killed the dinosaurs. But large enough to destroy most of the life on any continent it struck. And to affect life on Earth forever. It missed us by a mere six hours. Six hours in almost five billion years? Both events ensured that mankind would survive.

Was God involved?

> Life as a dice game is one conclusion. But remember, the mammals made it through the disaster 65 million years ago. It was the dinosaurs that lost the race. Perhaps, instead of fortuity, we have discovered a cosmic tuning to the flow of life. Rather than chance, the fossils that mark the demise of the dinosaurs may be evidence for a teleology.
>
> Schroeder, Gerald L. The Science of God: The Convergence of Scientific and Biblical Wisdom.

Summary:

> *The Torah, and the chapter we call Genesis, was written 3500 years ago. During that time, the original Hebrew language changed many times.*
>
> *Remember, when written, it was for an ancient, mainly nomadic illiterate people. Things are not always clear in today's language.*
>
> *So yes, dinosaurs were mentioned in the Bible. But to discover that takes some digging.*
>
> *Is there meaning to our lives that transcends the splendour we see about us? Something that goes beyond the physical? And if*

so, how can we probe that meaning? So much of life seems to hang upon uncontrollable events.

We will explore those questions and more in Part Two.

PART 2
THE STORY SO FAR

So you speak to me of sadness,
And the coming of the winter
Fear that is within you now,
That seems to never end.

And the dreams that have escaped you
And the hope that you've forgotten
You tell me that you need me now.
You want to be my friend.

And you wonder where we're going.
Where's the rhyme, and where's the reason
And it's you cannot accept
It is here we must begin.
To seek the wisdom of the children
And the graceful way of flowers in the wind

Though the cities start to crumble
And the towers fall around us.
The sun is slowly fading,

Michael Chambers

and it's colder than the sea

It is written from the desert
To the mountains, they shall lead us.
By the hand and by the heart
They will comfort you and me.
In their innocence and trusting
They will teach us to be free

And the song that I am singing
Is a prayer to non-believers
Who come and stand beside us,
We can find a better way.

From a song by John Denver, 1969

CHAPTER 16

The Story So Far (… it's all about religion, stupid!)

In the first part of this book, we looked at the question: Where are we from? We used science and ancient texts to explore that question and establish that we are "…fearfully and wonderfully made." With great reverence, we are unique and set apart.

> The word "fearfully" is translated from Hebrew to mean "with great reverence," "heartfelt interest," or "respect." The word "wonderfully" is translated from Hebrew to mean "unique" or "set apart." In this context, "fearful" does not mean to be literally scared or afraid.
>
> Psalms 139:14-16. New International Version.

Next, we will explore the questions: Why are we here? Where are we going? Has mankind made progress? What is our destiny?

But before we explore some of the more extraordinary aspects of this luckiest of all planets, we need to examine the downside. What I call the three-legged Stool of Misery that humanity has learned to inflict on itself: Slavery, War, and Discrimination.

As I explained in my introduction, "…understanding the world requires more than scientists and theologians. It needs philosophers, poets, painters, and people who dream and envision." So, let's leave the world of science for a while and look at the world around us through the realm of poets and dreamers.

A few months ago, as I was writing the closing chapters of part one, I was stricken by a severe infection that led to heart failure. I died many times in one night, saved by an excellent medical team. After six weeks in hospital, I am now well again and thriving. More about that story later. Needless to say, that experience has had a significant effect on my writing.

In the world of navigation, from the ancient mariners to modernity, it is necessary to know where we have been to establish where we are and where we are going. On our journey to understand why we live on The Luckiest Planet, we need to understand the past. What can we learn from the past to make this planet a better place? This is not a history book, so I will share only those events that have shaped my experience and impacted that of my family before me. Then, we will look at the future through the prism of the past. A future filled with hope and optimism!

Many would call me "old fashioned." It is probably meant in a derogatory way. But I say *I am* old-fashioned. My mind and thoughts have been fashioned over a long period by events I have experienced. So here is my story, so far

***

The Scourge of Slavery

It would be difficult to tell the story so far without examining the most significant stain on all of human history – slavery. Many think it is a notorious practice that only happened between Africa and America. That's true, but that's only a tiny part of the tale. Slavery has been a blight on every nation in the world at some time or another. It was practiced by every great Empire; it left a shadow over

The Luckiest Planet

every continent except Australia. Even today, there are tens of millions enslaved, a blot on our so-called modern societies.

Slavery was widespread at the time of Exodus, 3500 years ago. Indeed, the Hebrews were slaves to Egypt before they gained their freedom. The great empires of Assyria, Babylon, Persia, Greece, Rome, and the Ottomans, Mongols, and Chinese perpetuated the practice as a vital part of their economic system. Without slavery, their whole society and civilisation would collapse. And it did just that! Rome refined slavery into an art form.

One of the most moving stories I have ever heard is told by bestselling author and historian Tom Holland. He speaks of the horrifying and widespread Roman practice of consigning newborn babies to the rubbish dump.

> ... you have this extraordinary system of people going around the rubbish tips of various (Roman) cities looking for the weakest and humblest of all, which are abandoned babies. You know, that girls might be picked up by pimps or slave dealers and raised to be prostituted, but most likely, they would be flushed down into the drains. Archaeologists (today) can generally tell that the pre-Christian drainage system was littered with babies' bones; post-Christian, there are no bones.
>
> Tom Holland: How The Christian Revolution Remade the World.

That is a remarkable and revealing fact. He explains that the early civilisations were barbaric and immoral. For instance, the Roman view of sexuality was so different from ours that it is hard for us to comprehend. Roman culture was divided socially into the haves and the have-nots. For men at the higher end of society, sex was simply a normal bodily function to be performed on demand with whoever was at hand or took their fancy. Indeed, the word for urination was the same word used for ejaculation. So, in biblical times, Roman elites considered availability, not gender, when wanting sex. Whether male or female was irrelevant, slaves had no choice.

Within an amazingly short period, the radical Jewish religion of Christianity challenged the 1000-year Roman Empire and won. With humility, kindness, and compassion!

So, what does that have to do with me? Quite a lot. I was 10 years old and in England for a time. My grandfather had a travelling role as an Elder. I regularly attended the weekly meetings of the Society of Friends (Quakers) in Sunderland, Middlesborough, Newcastle, and Leeds.

Started in the 1600s by George Fox as a protest against corruption in the organised church, both Catholic and Protestant; they were so named because they practiced "shaking" whilst praying. In the 17th, 18th, and 19th centuries, they were a political force to be reckoned with, a radical group of fundamentalists resulting from the increasing biblical literacy sweeping England. George Fox taught of the priesthood of all believers, that no man was above them, and that there was no need for a priest, vicar, or anyone *to be over* the congregation. He also taught we are all made in God's image, that there is some of God in us all, that we are all called to communicate with God directly, and that killing and war were wrong; most Quakers were pacifists. He was jailed for his beliefs, and many followers were imprisoned. Some were executed. A meeting of The Friends was characterised by complete silence and prayer to examine the Spirit within. It was broken only occasionally and sometimes not at all by someone feeling it proper to share out loud.

They were persecuted from the start, so much so that the colony of Pennsylvania was founded by William Penn in 1682 as a safe place for Quakers to live and practice their faith in the New World.

In the meetings that I attended, mainly with my grandparents, I was taught how the Quakers were instrumental in banning slavery, not only in the Americas but in South America, Europe, and the Islamic countries of Africa and Asia. The story is not well known, but who lit the flame?

It starts, as many great stories do, in a very small way —by a pair of dwarves who also happened to be Quakers! Persecuted for their faith and stature, they left England for greener pastures. In Dominion: The Making of the Western Mind Tom Holland explains.

> To cross the Atlantic, then, was to lay claim to the liberty that Paul had proclaimed to be every Christian's. 'It is for freedom that Christ has set us free.' In the autumn of 1718, when a Quaker named Benjamin Lay sailed for the Caribbean with his wife, Sarah, he could do so confident that they would literally be among Friends. Barbados, an English colony that had existed for almost a Century, now belonged to a British Empire following England's union with Scotland in 1707. In the words of one settler, it was 'a Babel of all Nations and Conditions of men'. Yet even amid the colour and clamour of Bridgetown, the island's principal port, the Lays stood out. Both were hunchbacks; both were barely four feet tall.

There, amongst their own faith, they were horrified to find a society not far removed from the ancient Romans in their attitude to slavery, particularly black people.

The islands of the Caribbean first came to European attention when Christopher Columbus arrived on his second voyage in 1494. His fleet of twenty ships carried enough supplies, weapons, and ammunition to subdue the indigenous inhabitants. As with nearly all European settlements, the local population was soon devasted by diseases brought by the settlers. But the islands supplied what Spain required; gold in the mountains, sugar cane in the fields, and a population to convert, mainly by the sword, to the Catholic faith. The devastation was complete; they progressively conquered Panama, most of Mexico, and much of South America. In a treaty with neighbouring Portugal, the two dominant seafaring nations divided the world; Spain took the west and the Americas. Portugal took the East and the Spice Trade. Within a century, nutmeg would be worth more in Europe than gold! But the Americas provided sugar, and many ancient civilisations and riches abounded. Both empires were founded on the practice of conversion to Catholicism – or face death.

So, when Benjamin and his wife Sarah arrived in Barbados, they found an economy based on the production of sugar and black slaves imported from Africa, supplied in the main by other black Africans! Slavery has always been prominent in Africa, particularly among the Arab nations.

The indigenous population had been virtually wiped out, and the dark-skinned Africans were judged to be more suited to hard labour in the tropics. They were bought and sold as chattels, living in horrendous conditions, and justified by the Christian community because they were saving souls. For instance, the punishment of a runaway might be viewed as doing God's work! Both the Old and New Testaments of the Bible can be interpreted to favour slavery.

But Benjamin and Sarah realised that they were in error. The two little people were horrified and waged a war of attrition against the whole Christian community. They purchased only products not grown or produced by black hands. They grew as much food as they could, growing weak in the process. They travelled from meeting to meeting, preaching the end of slavery, being outlawed and ostracised by many. But over time, they won the population over. Tom Holland explains

> To trade in slaves, to separate them from their children, to whip and rack and roast them, to starve them, to work them to death, to care nothing for the mixing into raw sugar of their 'Limbs, Bowels and Excrements,' was not to be a Christian, but to be worse than the Devil himself. The more that the Lays, opening their home and their table to starving slaves, learned about slavery, the more furiously they denounced it – and the more unpopular they became. Forced to beat a retreat from Barbados in 1720, they could never escape the shadow of its horrors. For the rest of their lives, their campaign to abolish slavery – quixotic though it seemed – was to be their pilgrims' progress."

Dominion: The Making of the Western Mind

It would be exciting to report that the movement started by Benjamin and Sarah Lay spread like wildfire. But that was not the Quaker way. They indulged more in consensus and bridgebuilding, as well as in argument and debate. Nevertheless, by 1787, a Yorkshire member of Parliament, William Wilberforce, came to the same conclusions the two dwarfs had reached – slavery was against God's Word. A recently converted evangelical Anglican, Wilberforce cobbled together a coalition of abolitionists; Quakers, Protestants, Anglicans, Catholics, and many non-believers who were convicted by the righteousness of the cause. It took 20 years, but the Slave Trade Act of 1807 was passed. It was the abiding passion of William Wilberforce's life; the abolition of slavery. In ill health, he was not only instrumental in that achievement; he also founded the RCPCA!

From then on, Britain was on a holy mission; to abolish slavery worldwide. It unleashed its powerful Navy on the slave trade worldwide, often directly against its own economic interests. Regardless of nationality, it attacked every slaver operating in the Atlantic and on the way with their cargos of misery to Brazil and the New World.

Tom Holland concludes:

> And so it was, in the midst of a deadly struggle for survival against Napoleon, that the British parliament had passed the Act for the Abolition of the Slave Trade, and in 1814, that Lord Castlereagh faced across the negotiating table by uncomprehending foreign princes, had found himself obliged to negotiate for the eradication of a business that other nations still took for granted. Amazing Grace[1] indeed... increasingly, it was in the language of human rights that Europe would proclaim its values to the world."

Dominion: The Making of the Western Mind.

---

1 Amazing Grace. The Movie. 20th Century Studios. 2006.

The Bible has quite a lot to say about slavery. It was common practice in antiquity. Noah predicted that Caanan would go into slavery, and that was fulfilled by the Hebrews in Egypt; they experienced horror and bondage. But the Lord set them free and bought them out of Egypt with a mighty hand. Paul's letter to the Galatians in the New Testament seems to light the way. Galatians 5:1 says

"It is for Freedom that Christ has set us free. Stand firm, then, and do not let yourselves be burdened again by a yoke of slavery.

The Quakers argued that "there is that of God in every man," referring to Genesis 1:27

Therefore, how was it possible to keep a man or woman in bondage who were made in God's image?

I am not a Quaker – I did not follow that path in life. But I am so proud of my forebear's bravery, tenacity and wisdom. It is a story worth telling. I salute them.

Summary:

Slavery has always been a part of the human condition, sometimes an important and irreplaceable part of a nation's economy.

It was only through the growing influence and understanding of the Bible that people came to realise that slavery was evil, and not part of God's plan for humanity.

That's when things started to change.

CHAPTER 17

The Horror of War

Folk singer Bob Dylan wrote
Come you masters of war
You that build the big guns
You that build the death planes
You that build all the bombs
You that hide behind walls
You that hide behind desks
I just want you to know
I can see through your masks.

We have discussed the morality of slavery. Now, we must turn our attention to that other curse of humanity: war. It is a complex subject to write about, read, and understand. But to know where we are, we must know where we have been. To realise our destiny, we must understand our past and our history.

My maternal grandfather was an interesting fellow. Charles Townsend Brown was known to all as Charlie Brown! Considering how little time I got to spend with him, he has had a disproportionate effect on my life. I was born into his home in Yorkshire and lived there till age five, when, together with my younger sister, Trisha, we migrated to New Zealand. Then we spent another year with my grandparents when I was aged ten. By then, they had moved to Whitburn, a small northeast coastal village near Sunderland in County Durham.

Charlie was a quiet, clever, intelligent man. A devout Quaker, as was my grandmother Phoebe; nonetheless, he was fun to be around. He lived on Adolphus Street, with rows of adjoining houses on one side and a farm on the other. Yes, that's right, a farm; this was England in the fifties. The village farm just over our fence was all concrete and stone, with not a blade of grass to be seen. With cows, pigs, poultry, and plenty of manure, the animals wandered freely, as I would discover when I went to retrieve my lost soccer ball when playing in our (concrete) backyard with my grandad.

The house was small for two families. Two bedrooms, a small sitting room, a bathroom, and a scullery (kitchen) that led to the back street, where they delivered coal down a shute and into the bin in the scullery. My recollection is we had a fire most of the year. He taught me to play billiards on the kitchen table. He helped me build a crystal radio from scratch that I used to listen to short-wave radio enthusiasts from around the world.

The only thing resembling a garden was a small patch of lawn at the front, about ten feet square, that Charlie kept trimmed. And here was his pride and joy – a pair of beehives. He gave one of the hives to me, and we practised the noble art of beekeeping for a year together, culminating in a prize at the Yorkshire Show for my honey.

Charlie drove a Jowett, a small black four-door car made in Yorkshire that both families squeezed into; just! We had many memorable trips across the north, including Hadrian's Wall, Bolton Abbey, and Durham Cathedral, where King Richard the Lionheart was buried. My grandfather was a professional engineer and a travelling representative for Simmonds Nuts. It was founded in 1936 by Sir Oliver Simmonds and produced a range of aerospace nuts and accessories. More recently, it was famous for inventing the Nyloc Locking Nut System. I would often accompany him on his business trips across northern England. I'm sure we had many long and interesting conversations, but only one has stayed with me. It probably answered my question: "How did you lose your finger, grandad?" Then, for the first and only time, he spoke of the Great War, what we call the First World War.

Charlie Brown was a conscientious objector – that is, he believed killing was wrong and forbidden by the Bible. More about that later. He was a very young man but also a patriot and a proud Englishman. When the war came to England, he volunteered as a medic. He was quickly trained and shipped off to the battlefields of France, which, by every account, was a living hell. I recall that he was at the Battle of the Somme, which was late in the war, so he had been incredibly fortunate to avoid death or injury. The ambulances were often horse-drawn affairs, pulling covered wagons with a prominent Red Cross, best depicted in that excellent 2011 Spielberg movie Warhorse. We frequently talk about the bravery of soldiers in wartime. What he did called for a unique courage, rescuing the dead and wounded as the battles raged. If he was affected by the horror he must have experienced, like many returning soldiers, he never mentioned it.

A shell, not quite close enough to kill him, tossed him into the remains of a tree. While he was there, he was gassed – both sides used a green-coloured chlorine gas that was often fatal. Because he was up the tree, he missed most of it as it crept along the ground

The Luckiest Planet

below. Miraculously, his primary injury was to lose the index figure of his right hand, though, for the rest of his life, he could still feel it as though it was there. He explained that when the war was over, and he returned to umpiring cricket, he would indicate a batsman was out by raising his index finger – and everyone laughed when nothing happened except a raised fist!

And that's all I know about Charlie Brown's war. He was a good man.

* * *

Since the time of Adam, when Cain killed Abel in a fit of envy because of a disagreement over a sacrifice, so have millions upon millions of men and women been sacrificed on the altar of war. That singular murder was the beginning of brother against brother, town against town, and nations and empires engaging in mortal combat. Was it wrong? Absolutely – like slavery, a blot on the human character. Was it always wrong? Well, no. Like the Reality of Duality, the moral imperatives of a just war were established, starting with the early church fathers and culminating in the 20th Century. There rose the concept of a "just" war. A conflict waged for the common good, in good conscience toward God.

Before 1914, Europe was a cot case. For a thousand years, the fate of empires had ebbed and flowed along with the borders. War was seen as an essential, honourable, and patriotic endeavour, generally invoking God and most often seeing Him as being on each side. As technology improved and more and better ways were devised to kill, the number of soldiers required on each side increased exponentially. A new phrase was introduced: the generals used men as "cannon fodder." Once numbered in the hundreds, battles now required soldiers in the millions to fight trench-to-trench and face-to-face.

They called it "The War to End All Wars," and many believed it. The carnage was horrific, surpassing by an order of magnitude anything thus far in history. Ever! My grandfather was one of the fortunate ones; 38 million were not. The total number of casualties in World War I is estimated to be around 40 million, with about 15–22 million deaths and 23 million wounded soldiers. It resulted in the demise of the Russian, Austrian, and Ottoman Empires and the beginning of the end of the British Empire. A worldwide flu epidemic resulted in over 500 million deaths, probably related to the war.

Unlike the second world war, the bloodbath of 1914-1918 was not a just war. It was a savage industrial slaughter perpetrated by predatory imperial powers, locked in a deadly struggle to capture and carve up territories, markets, and resources. The crowned heads of the European nations were closely related and shared the same God and Bible. But they still went to war.

Germany was challenging the powers of Europe, seeking its own Empire. The excuse was a diplomatic struggle in the Balkans, an assassination of one of the ruling families, as Austria, Hungary, and Russia fought over the crumbling remains of the Islamic Ottoman Empire. The elites of Europe, oblivious to the human cost, went to war and shared the blame for the murderous barbarism they created.

There is one rule concerning a just war – morality and justice are always decided by the victor. It has always been thus!

Previously, we learned how a faulty, defective gene, while generally harmful, can benefit a live organism. Likewise, beneficial effects were flowing from the Great War. One resulted from actions by the Australian Forces in the Middle East.

From a population of less than 5 million population, about 420,000 Australians were listed for service in the Great War. That was nearly forty per cent of the male population aged 18-44! Of those, 215,000 were either killed or wounded. That was our contribution to the Empire. Not as well-known as Gallipoli was the contribution of the Australians to the demise of the Ottoman Empire at Beersheba in Palestine. In my book Walking Among the Stars, I tell the story of that battle and explain the significance, circumstances, and battles leading up to it.

> The Charge of the Light Horse at Beersheba in 1917 had been widely publicised in Australia. It was one of the most incredible cavalry charges of all time. Indeed, in modern times, it compared scope and bravery to anything that Genghis Khan's Mongols, Alexander, Hannibal, or Napoleon had accomplished. In fact, Australian Prime Minister Billy Hughes declared, "In the history of the world, there has never been a greater victory than that which was achieved in Palestine!
>
> Michael Chambers. Walking Among the Stars: An Epic Tale of the Australian Outback

The victory was the first significant loss the Ottoman Empire had suffered in the war and ultimately led to their defeat.

In Dominion: The Making of the Western Mind Tom Holland explains:

> In Palestine, the British won a crushing victory at Armageddon and took Jerusalem from the Turks. In London, the Foreign

Secretary issued a declaration supporting the establishment in the Holy Land of a Jewish homeland.

Jews soon began flocking to their restored home.

As a result of the Great War, the League of Nations was established. This paved the way for the United Nations in 1945: at least they were talking to each other. In 1948, Israel was declared a nation by the United Nations. The beneficial effect of a faulty gene!

Now, over 100 years on the Great War is only wasted if we allow it - if we learn nothing from it. That would be a waste. But we did learn small lessons, and at a terrible cost. But that is progress of a kind. And progress is not worthless. But it has not answered the most challenging question– what are we doing here? Do we have intrinsic value as human beings? Or is war the final mediator?

***

If at first you don't succeed – try again!

Peace lasted a mere two decades. My parents took part in what we now call the Second World War. Still in their teens, my father was a Spitfire mechanic in the RAF, one of those who kept planes flying during the Battle of Britain. My mother was a WREN (Women's Royal English Navy.) She was a secretary to some high-up Admiral in the north of England. As far as we know, my parents met after the war.

We were the first of the Baby Boomers, the generation of post-war babies. Eight years after the war, we emigrated to New Zealand. Although very young then, I still remember Churchill's voice over the radio. I was still relatively young when I read Churchill's

"History of the Second World War," all six volumes! It traced the history of Europe from the end of the Great War to the end of WW2.

The victors made a mess of the initial peace treaty with a defeated Germany. The terms of surrender were so terrible as to almost guarantee another European war – it only took an Adolf Hitler to start it. A lowly Corporal in the first war, a tragic figure of evil and violence, he was buoyed and supported by a population mired in defeat and poverty. All requisites for a good war! My mother was a stamp collector in those brief years of peace. I still have the album – hundreds of stamps from Germany, each worth thousands of Deutschmarks as the German economy crumbled. Hitler promised better times: a thousand-year Reich, a glorious worldwide empire.

It was a tantalising dream, and most German people bought into it.

It was also a flawed dream and an even more expensive lesson for humanity. Military *deaths* from all causes totalled 21–25 million, including *deaths* in captivity of about 5 million prisoners of *war*. About 10 million others, including nearly 6 million Jews, plus Gypsies, many minorities, and other "non-desirables." It is estimated that there were, in total, about 85 million casualties; most were from the civilian populations of China and the Soviet Union.

The man Churchill described as the most powerful man in history, General Dwight Eisenhower, was the Supreme Commander of the Allied response in Europe. A future US President, he was a very humane man. Yet he sent over 160,000 troops into battle on the beaches and fields of Normandy, France, knowing many would die – and they did!

After the war, Dwight Eisenhower said this:

Every gun that is made, every warship launched, every rocket fired signifies, in the final sense, a theft from those who hunger and are not fed, those who are cold and are not clothed. This world in arms is not spending money alone. It is spending the sweat of its labourers, the genius of its scientists, the hopes of its children. This is not a way of life at all in any true sense. Under the clouds of war, humanity is hanging on a cross of iron.

Dwight D. Eisenhower. Chance for Peace speech. Washington, April 16th 1953

Eisenhower did not enjoy war!

When I was young, amid the nuclear bomb madness, we were taught that at the sound of the alarm, we were to duck under our desks with our heads down. Fat lot of good that would do. One bright spark suggested, "When you see a blue flash, put your head between your legs and kiss your ass goodbye!"

I don't like war either.

***

The Courage of Tommy Brown

There are thousands, no, probably millions of stories of sacrifice and bravery in WW2. There are none so brave that had such massive consequences on the war's outcome than that of Thomas William Brown.

1942, the war raged, and at fifteen, like many other young men (I hesitate to call him a boy), he lied about his age and joined the Royal Navy. He was assigned to "HMS Petard" as a cook in the

The Luckiest Planet

enlisted men's canteen. The P Class destroyer was sent to patrol the Mediterranean near the coast of Egypt.

There were reports of a German submarine and three other vessels being sent to investigate. After ten hours of depth charging, U-559 was forced to surface. Under fire from "Petard," the crew opened the seacocks to sink the submarine and abandon the ship.

Together with Francis Fasson and Colin Grazier, Tommy and his mates stripped off, rowed a whaler to the sinking sub, and swam naked to the mortally wounded vessel. They entered through the conning tower and began a search for anything of value. Finding a set of keys in the captain's cabin, they unlocked the desk drawer to find two code books: the Short Weather Cipher and Short Signal Book. With the precious books in one hand, Tommy carried these up the iron ladder to the conning tower and swam with them back to the whaler. After his third trip carrying documents, he swam back to the submarine, crying out to his shipmates, but it was too late. They had drowned. He rowed sorrowfully back to his ship alone.

Decades after the war, it was discovered that the genius Alan Turing and the team of code-breakers at Bletchley Park had used Tommy's books to crack the German Enigma Machine. For the rest of the war, hundreds of thousands of lives were saved, and the course of the war changed.

"Allied convoys in the Atlantic could be directed away from known U-boat locations. Winston Churchill wrote that "the actions of the crew of *Petard* were crucial to the outcome of the war."

Tommy, the hero, was not discharged but sent back to North Shields because of his age. He later rejoined the Navy, and while on shore leave, his home caught fire. He died while trying to rescue his youngest sister, Maureen. He died unaware of his impact on the war and human civilisation!

He was the youngest person to have received the George Medal for bravery.[2]

Tommy Brown was born and lived in North Shields. I have been to his village, on the mouth of the River Tyne. My grandfather, Charlie Brown, lived only a few minutes away. There is a whole community of 'Browns' in that part of the world. Are we related? I would like to think so

***

A Just War is Still War.

War is like abortion. It should be rare but available only in a unique set of circumstances. And like abortion, establishing a legal framework for a just war is brutal. But I believe both are wrong - except in exceptional circumstances. And not necessarily because the name of God is invoked or called upon.

The fields of Europe and Asia, including Israel, are littered with the bones of warriors who have fallen in wars over religion. The numerous crusades over many centuries to free Palestine and Jerusalem pitted Christian against Moslem. The expansion of the Ottoman Empire set Moslem against Moslem. When Martin Luther pinned his demands to the door of a Catholic church, it sparked civil wars, Catholic against Protestant, father against son, brother against brother, in England, France and Germany. In recent years the world had been rocked by a number of wars and

---

[2] Wikipedia: Tommy_Brown_(NAAFI_assistant)

terrorist related incidents, radical Islam against a west influenced by Christian thought and culture.

"Conflict took place every year of the 20th Century; the world was free from the violence caused by war for very short periods. It has been estimated that 187 million people died as a result of local wars from 1900 to the present. The actual number is likely far higher."[3]

And yet, since 1945 and the dropping of two atomic bombs on Japan, we have had a period of relative freedom from a *major* war. Many of those that occurred could more correctly be called unjust wars. Perhaps the most unjust of all was the war perpetrated by Ethiopia and Eritrea in 1998.

Ethiopia is an ancient land mentioned many times in the Bible as the land of Cush. In the latter part of the 20th Century, a civil war broke out – the rebels won. Two states were created; Eritrea in the north and Ethiopia to the south. More about Ethiopia in my last book, Makarrata.

An article in The Independent: There are no winners in this insane war explains.

> Profound differences began to emerge. They adopted very different political systems. Ethiopia opted for an ethnic-based democracy. Eritrea remained a one-party state. Ethiopia became landlocked and dependent for trade on Eritrean-run ports, but the customs and tax systems diverged. Citizenship was never sorted out, nor was economic policy. When Eritrea introduced its own currency in 1997, Ethiopia suddenly said it would not accept it. Ethiopia accused Eritreans of smuggling

---

[3] Imperial War Museum: Timeline of 20th and 21st Centuries.

coffee and other Ethiopian produce across the border. And that border itself was never properly delineated.

But this was a land of famine, feuds, and poverty. The festering hatred between the two countries became an all-out war in 1998. The supposed issue was the border; more accurately, the town of Badme had a population of 1500 people!

The war lasted for two years. Upwards of 300,000 people died in combat, not to mention the millions of displaced refugees. When an awkward peace broke out, the border remained essentially unchanged.

This was not a just war.

It can be argued that the American Civil War over slavery *was* a just war. Despite the nearly 850,000 killed in battle, the value of slaves in southern states exceeded the total value of all other land, treasure, and wealth. The Quakers were so invested in the abolition of slavery that they declared it a just war, and many of their young men took part in the fighting!

It can be argued that the Vietnam, Afghanistan, and Iraq wars were not just. So, what's the difference? I cannot find any justice in the war in Ukraine – it seems to have no end for little gain. Contrary to that, I believe Israel has the right to defend itself against Hamas, a mortal enemy.

A "Just War" has four elements or requirements. One: Is it for a cause? Perhaps self-defence against an aggressor. (The schoolyard bully springs to mind.)

Is it by a legitimate authority? i.e., a national government? Is it proportional? That is, does the punishment fit the crime?

And finally, but perhaps most important? Is there the slightest chance of more talk or diplomacy?

On the other hand, war is wrong if it seeks revenge, displacement of people, or territorial gains. World War 2 is widely regarded as being just and having beneficial outcomes. For example, as well as an Allied victory, China and most of Asia became free of Japanese influence. The axis of Germany-Italy-Japan was broken, and the United Nations was formed.

But I would add one last element to think about just war. There must be a "Just Peace." That did not happen in the Great War. Without a just peace, the wounds and disagreements that caused the war will continue to fester. It happened after WW2, when the Western powers, mainly the USA, ensured the growth and reconstruction of both Germany and Japan; they are now thriving democracies. That was just and created a peace that continues to this day.

Another restraint on starting a fight, a battle, or a war stem from a basic understanding of human nature. It is called deterrence, or peace through strength. This doctrine has been successful throughout the Cold War years and recently. Simply: common sense! In Luke 11:21-22 CJB

Jesus said:

> When a strong man who is fully equipped for battle guards his own house, his possessions are secure. But when someone stronger attacks and defeats him, he carries off all the armour and weaponry on which the man was depending, and divides up the spoils.

The Old Testament is replete with war. In fact, it states clearly there is:

> a time to love and a time to hate,
>
> a time for war and a time for peace.
>
> Ecclesiastes 3:8 CJB

Speaking of wars, just or otherwise, Jesus said.

> "You will hear the noise of wars nearby and the news of wars far off; see to it that you don't become frightened. Such things must happen, but the end is yet to come.
>
> Matthew 24:6 CJB

Yet like its twin evil, slavery, war for the wrong reasons is never condoned but is often depicted as just. Again. The duality of reality. Yet our path must always favour Jesus' words. Matthew 5:38-40 CJB

> You have heard that our fathers were told, 'Eye for eye and tooth for tooth.' But I tell you not to stand up against someone who does you wrong. On the contrary, if someone hits you on the right cheek, let him hit you on the left cheek, too! If someone wants to sue you for your shirt, let him have your coat as well!

It would seem that the right way, the better story, is for love, acceptance, and forgiveness. The Jesus way. That appeasement, often vilified in modern times, is the authentic way of reconciliation. In Luke 17:3-4 He says.

Watch yourselves! If your brother sins, rebuke him; and if he repents, forgive him. Also, if seven times in one day he sins against you, and seven times he comes to you and says, 'I repent,' you are to forgive him.

To finish, let's consider the words of Albert Einsteins observation about war. "I know not with what weapons World War 3 will be fought, but World War 4 will be fought with sticks and stones."

***

CHAPTER 18

The Stupidity of Racism!

Slavery and war are two legs of the Human Misery stool – that is, the wretchedness imposed by humans upon humans. The third is discrimination, both racially and in many other ways. In my introduction, I told of my first encounter with apartheid, that is, organised racism, in South Africa when I was young. Nearly forty years later, I was subjected to another version – in Egypt at the other end of the African continent!

On a Sunday afternoon in September 1994, I was being held at gunpoint, along with about 25 friends, in a dungeon beneath Cairo Airport!

That was a bit careless. How did that happen?

In the late nineties, I gave up a decade-long career with BHP. I loaded my wife, five kids, furniture, and dog into a truck and minibus headed for Darwin. It was a family adventure that would last more than twenty years. A couple of years later, I was involved in a ministry of reconciliation and bible teaching amongst Aboriginal people, mainly on Elcho Island, the northernmost tip of Arnhem Land.

The Yolngu are wonderful, welcoming, cheerful people, and I find them quite different from other cultures. Probably due to their proximity to the ocean and centuries of contact with the outside world, mainly southeast Asia, but as far away as China! In my latest book, Makarrata, I explore their story, a historical fiction spanning thousands of years, culminating with Western civilisation's arrival as Methodist missionaries in 1922.

We were a diverse mob – I and one other were white, though he was a redhead! He had spent most of his life with the Yolngu, spoke their various languages (there were many), understood their cultural ways, and was married to a prominent clan leader. Their three children were brown – the rest of the group were black, jet black. It didn't worry us – though it would worry some. They had asked me to organise and lead a trip to Israel, which was a wonderful learning experience.

Getting the passports was a challenge – they all had three names; an English name, a Yolngu given name, and a clan name. Think thirty letters. When I handed in the passport applications three months early, they were shocked and dedicated an official to the project. Just as well, the passports arrived in Darwin only a week before departure. And by the way – they each paid their own fares.

The cheapest and shortest way was through Singapore, Dubai, and Cairo. An overnight stay, and on to Israel the following day.

We had a great trip, sightseeing in Singapore for a few days. We landed in Cairo late in the afternoon. That's when the trouble started. We had been forewarned about the practice of "baksheesh." As the group's banker, I had set aside an appropriate amount of cash. Wikipedia says:

> Baksheesh is given as alms: By praising Allah, beggars make it possible for their fellow men to serve Allah by giving baksheesh. - Baksheesh is given for services rendered: Egyptians give baksheesh to provide service. This is the closest to tipping, as Westerners know it.

Obviously, we were not welcome; we were spat at, sworn at and subjected to many dark looks. The "baksheesh" was extortionate, far more than we expected. Finally, we made it to the waiting bus, an old, antiquated affair, with our lives, luggage, and yidaki (a didgeridoo,) too long to fit in luggage, and anxious to be out of there.

We spent the night at the Anglican hostel – they were terrific. They were hosting a conference of bishops from right across Africa. I'm sure Desmond Tutu was there, but I don't remember him. When they found out we were there, they invited us to dance and sing, which we did, with white painted faces, yidaki, and bilma (clapsticks.) The bishops were delighted, and we all had a wonderful evening. Later that night, we discussed what happened and why everyone had been so antagonistic at the airport. Hateful, even. We concluded that a) They didn't like black people. b) They didn't like white people mixing with black people. And c) They didn't like white people or black people going to Israel!

We arrived at the airport early the following day, ready for the short flight to Israel. Apparently, that was not a good idea. At the ticket counter, we were met by an armed squad with a major in charge. He confiscated our passports and surrounded by machine guns, we were escorted downstairs to a large, ill-lit room with no windows, no chairs, and a toilet in one corner that had not been cleaned for some time. Armed guards at the door.

The Major led me to another, smaller room with two seats and a table with all our passports on it. He seemed angry. He interrogated me for about half an hour.

"Where are you from?"

"Arnhemland, Australia."

"Who are you?"

"I'm a Christian missionary. These are my friends. English is their second language."

He wasn't happy. "Why are you going to Israel?"

"To dance and sing at the Feast of Tabernacles. There will be thousands from across the world there."

There were many more questions along the same line. Then, I rejoined the others. It's not so much we were frightened, which we were, but I was worried about missing our flight.

The hours passed. It was stinking hot, with no ventilation. We spent the time singing, talking, praying, and laughing. These were Yolngu people; that's what they did. Finally, the guards took us upstairs to the main waiting area – and left us, just like that.

The flight, though delayed, was still showing on the screens. I asked them to follow me, and I sprinted down the corridor to the gate. At the entrance were two guards armed with machine guns. By this time, my blood was up – and to their surprise, I forced my way past them into the departure lounge. And there, thrown carelessly on the floor, were our passports! All of them. By this time, the others had arrived, breathless and excited. Our plane was on the tarmac, with all our luggage piled next to the aircraft on the tarmac. I knew it was ours because the yidaki was sitting on top.

A worker helped load our luggage, and we boarded the plane and found our seats. There were a few others, and we were not asked for tickets or boarding passes! We all started singing and praising God when we took off for Israel.

The next day, in Jerusalem, I phoned the Australian Embassy. They were horrified and assured me things would be different on the return flight home in three weeks.

And they were. When we arrived at Cairo airport, we were met off the plane stairs by three Australian Embassy officials and an extensive smiling Egyptian official. We were whisked through Customs to cries of "baksheesh, baksheesh" from the people who had previously shunned us. The Egyptian official led us in a V formation, followed by our luggage trolley, passing out wads of money left and right as we passed through officialdom like Moses and the Hebrews passed through the Red Sea! And there was our coach, a modern air-conditioned tourist bus that quickly had us at the hostel.

The following day, bright and early, we left our accommodation to find the luxury coach, the embassy officials, and the grinning big Egyptian man waiting to repeat the performance for our departure from Egypt. And they did, with much wringing of hands and smiling faces. Thank you, Australia!

***

The third leg of the human stool of misery is racism and its attendant discrimination, such as that displayed by the Egyptians.

It may surprise you to know that a lot of this is a relatively recent phenomenon. For instance, there is no racism in the Bible – in fact, skin colour is never an issue. Perhaps the most outstanding example is when the Queen of Sheba (modern-day Ethiopia) travelled to Israel to meet King Solomon. There is controversy concerning her homeland, but she was undoubtedly darker-skinned than the King. But this is never mentioned. While we were in Israel, we met with an Ethiopian Jewish immigrant community – they were very dark-skinned.

Jesus taught about this:

> Stop judging by surface appearances, and judge the right way!
>
> John 7:24 CJB

The Bible explains how humanity should be viewed:

> After this, I looked, and there before me was a huge crowd, too large for anyone to count, from every nation, tribe, people and language.
>
> Revelation 7:9

So, what makes a nation. Language, tolerance, history and humour. These are the ingredients that make up the culture of a nation. Not skin colour!

The Bible teaches that we should view humanity as families, communities, tribes, nations, and language groups. But not one of these is dependent on appearances! The Bible is colour blind – so is God.

So, what is "race" and where did it come from? The Oxford Dictionary says:

> Race …the idea that people can be divided into different groups based on physical characteristics that they are perceived to share, such as skin colour, eye shape, etc., or the dividing of people in this way.

Race is an "idea" someone invented because they thought it would be beneficial. The National Museum of African American History and Culture explains:

> The term "race" was used infrequently before the 1500s to identify groups with a kinship or group connection. The modern-day use of the term "race' (identifying groups of people by physical traits, appearance, or characteristics) is a human invention.

Race divides human populations into groups, often based on physical appearance, social factors, and cultural backgrounds. It evolved separately, though with much the same roots and consequences, mainly though not limited to North America in the 16th Century and South Africa in the 17th Century. On both continents, the Bible was used to justify both slavery and racism (division by skin colour.) Contrary to Jesus' words, "Stop judging by surface appearances."[4]

***

How a Nation used the Bible to subjugate its people.

In the 1600s, Holland established a colony at the tip of South Africa, now Capetown. It provided a haven for shipping and safe passage to and from its enormous Empire in Southeast Asia, based on the lucrative Spice Trade, especially the nutmeg trade. (See my book Makarrata for more.) For the Dutch, Capetown was only the beginning of the exploration and subjugation of southern Africa.

---

[4] John 7:24 CJB

Many natives were forced into slave labour or killed as they expanded their domination into the outer regions.

The Afrikaan people, light-skinned Europeans, considered themselves a superior "race." Yet, they were a God-fearing people who suffered from a misinterpretation of the Bible. In short, they believed themselves to be the "New Israel" and not only appropriated God's promises for that nation to themselves but created a human hierarchy where their own "race" was on top, followed by "coloured" people and at the bottom "black' people, whom they believed to be barely human and should have been grateful to the settlers for bringing them the good news of the Gospel. The black nations of South Africa were devastated.

Fast forward to the middle of the 20th Century. In 1948, the very same year that Israel became a nation in its own land, South Africa was in turmoil; friction between the minority Afrikaans and English and the vast majority, black tribes of the hinterland. They ruled with an iron fist, aware that the slightest spark could ignite a bloody civil war. The government was harsh, bigoted, and cruel. For decades, the policy had been segregation, that is, physically separating blacks, coloureds, and whites. But then, with religious fervour, the ideology of "apartheid" was developed, a strange mix of religious theology and separatist politics. This was the horror of apartheid, meaning "apart" in Afrikaans. An article from South African History Online: A History of Apartheid, explains:

> The apartheid doctrine called for the separate development of the different racial groups in South Africa. On paper, it appeared to call for equal development and freedom of cultural expression, but its implementation made this impossible. Apartheid made laws that forced the different racial groups to

live and develop separately and grossly unequally, too. It tried to stop all interracial marriages and social integration between social groups. During apartheid, to have a friendship with someone of a different race generally brought suspicion upon you or worse. More than this, apartheid was a social system that severely disadvantaged the majority of the population simply because they did not share the skin colour of the rulers. Many were kept just above destitution because they were 'non-white.'

In 1994, when we were being held in Egypt, the horrific doctrine of apartheid was coming to an end. The African National Congress (ANC) won the elections in a landslide. Four years before that the African rebel leader Nelson Mandela was released from a brutal life sentence in jail, and became head of the ANC and led the fight against apartheid. He was inaugurated as South Africa's first democratically elected president in May 1994. Although apartheid was defeated, discrimination by blacks towards whites continues until this day. Which is sad.

Looking back on my time with the Yolngu people, it is hard to remember any incident where I felt I was being treated differently because of my colour. In fact, it would be difficult to find a people less prone to racial discrimination, probably with their centuries-long close association with the Macassans and other seaborn traders. But they were proud of their "blackness." They often performed a ceremony to enhance their colour and facial features at birth. I explain in Makarrata: The Australians of Arnhemland:

> Wonguma's second wife, took the baby and immediately pressed hard on its nose. This was to flatten it so that it would appear socially acceptable. Then she took a bag of black ash

and rubbed it all over the skin till it was very black. This baby would definitely have appropriately coloured skin. The right degree of blackness! Only then were the father and grandfather invited in.

* * *

The American Struggle for Human Rights

The most significant difference between racial relations in South Africa and America was that the Negros were in a minority. Even once slavery officially ended after the Civil War in 1865, the culture of discrimination lingered on. Though legally free, Negros continued to be subject to laws and cultural practices that kept many of them at the lowest level of society and denied them the vote. This continued for nearly two hundred years, even though they served with distinction in both World Wars, Korea and Vietnam. However, in the 1960s, everything changed when the American Civil Rights movement was born.

Martin Luther King Jr was born into a Southern Baptist household – his father was a preacher. King participated in and led marches protesting the right to vote, desegregation, and other civil rights. He was the iconic orator of the 20th Century along with Winston Churchill. I can vividly recall listening to his speeches as a young boy. Growing up in New Zealand, my friends were other Pakeha's (Europeans) Maoris and South Pacific islanders. We learned about racism, not so much by personal experience, but news from South Africa and America. So, when we heard his measured, strident tones delivering every word like a bombshell, we were excited as a group and a nation. Globally, there was a feeling that, as King put it, everyone "was free, free at last!" It was the beginning of something new and exciting. He said:

> I am happy to join you today in what will go down in history as the greatest demonstration for freedom in the history of our nation.
>
> Five score years ago, <u>a great American</u>, in whose symbolic shadow we stand today, signed the <u>Emancipation Proclamation</u>. This momentous decree came as a great beacon light of hope to millions of Negro slaves who had been seared in the flames of withering injustice. It came as a joyous daybreak to end the long night of their captivity.
>
> But one hundred years later, the Negro still is not free. One hundred years later, the life of the Negro is still sadly crippled by the manacles of segregation and the chains of discrimination. One hundred years later, the Negro lives on a lonely island of poverty in the midst of a vast ocean of material prosperity. One hundred years later, the Negro is still languished in the corners of American society and finds himself an exile in his own land. And so we've come here today to dramatise a shameful condition.

He says that he has come to cash a cheque; the Declaration of Independence was a promissory note to all Americans that they should all be free, black and white. Instead, this cheque has returned marked "declined – insufficient funds." So, on the part of all Americans, we've come here today to cash that cheque.

> Let us not wallow in the valley of despair, I say to you today, my friends.

And so, even though we face the difficulties of today and tomorrow, I still have a dream. It is a dream deeply rooted in the American dream.

I have a dream that one day this nation will rise up and live out the true meaning of its creed: "We hold these truths to be self-evident, that all men are created equal."

I have a dream that one day on the red hills of Georgia, the sons of former slaves and the sons of former slave owners will be able to sit down together at the table of brotherhood.

I have a dream that one day even the state of Mississippi, a state sweltering with the heat of injustice, sweltering with the heat of oppression, will be transformed into an oasis of freedom and justice.

I have a dream that my four little children will one day live in a nation where they will not be judged by the colour of their skin but by the content of their character.

I have a *dream* today!

...... And when this happens, and when we allow freedom ring, when we let it ring from every village and every hamlet, from every state and every city, we will be able to speed up that day when *all* of God's children, black men and white men, Jews, and Gentiles, Protestants, and Catholics, will be able to join hands and sing in the words of the old Negro spiritual:

Free at last! Free at last!

Thank God Almighty, we are free at last!

Martin Luther King's speech. Lincoln Memorial, Washington DC 1963

These words have echoed through the dark canyons of prejudice since – so my cry is – where is that dream. Why aren't we still living that dream today? What on earth could possibly have derailed that freedom train?

It is not yet lost, but its memory is become dimmed by the angry chorus of so-called progressives. Fuelled by fear, they shout.

> "Yes, there is such a thing as race. Yes, there must be differences. Otherwise, how shall we survive if there is no black or white, Jew nor gentile, slave or free? How shall we fuel the passions of our hatred of freedom?"

Yet, it is all based on a lie. The lie that the colour of a man's skin, or his features or disabilities, gives cause for fear and, therefore, they must be "separated" and treated differently. The lie that everyone must be divided into groups, not by tribes or nations or languages but by something created called "identity" groups.

Vivek Ramaswamy, a recent presidential candidate in the US says, "The best way to end racial discrimination is to end racial discrimination."

Have a look around you. Look at your family, friends, neighbours. Consider your town, your state, your country. Please don't stop; I want you to consider everyone. Don't stop – consider all who ever lived on this luckiest of all planets. Billions upon billions of people just like you – made in God's image - but different.

Do you know, are you aware, that no one looks like you? Can you believe that? You are unique. In all of history. No one has ever looked precisely like you. Consider that for a moment.

So why do we want to separate people based on *their differences?* We are all different. Let's celebrate that instead! There has been a trend in recent years to divide people with similar characteristics into groups, or "identities" for political reasons. This is the complete opposite of what Martin Luther King implored us to do. "Identity" politics is wrong.

"Stop judging by surface appearances, and judge the right way!"[5] said Jesus. Not by the colour of the skin but by the content of the character. How, oh how, have we missed that? The Biblical definition of sin is "to miss the mark." As a civilisation, how have we missed this mark?

The most heinous example of discrimination leading to racism happened in Germany during the 1930's. After WW1 Germany regained its position of the most educated and advanced of all European nations. But, once again, error crept in the back door, through the Church. Germany eagerly embraced the growing wave of Darwinism, understanding that man was merely an intelligent animal, descended from apes and monkeys, not made in the Image of God. Once a culture goes down this path, it becomes a slippery slope to hell. A particular interpretation of the Bible led to speculation that, indeed Jews, instead of being beloved of the God of Israel, were really vermin, to be exterminated by the superior Aryan master race. Like the Afrikaaners before them, much of the church in Germany applied the Biblical promises for Jew and for Israel to themselves. And German science "proved" that the Jews were indeed an "inferior" race. The error was catastrophic. With the tacit support of science and the church, Hitler's Nazis embarked on

---

[5] John 7:24 Complete Jewish Bible.

a dark, satanic crusade to stamp the Jews out from all the earth. And much of this well-educated well-read population watched on and did nothing. Six million Jews died, and many other minorities among them. Even today, and in spite of having their own homeland, the Jews worldwide number less than 15 million.

***

I was accompanied by a wonderful family group on my trip to Israel. He was very white; she was very black. Their beautiful children were various shades of brown. So, how are those put into separate "identities?" And when those children marry, some to black-skinned, some to white-skinned, how then are we to identify their offspring. Into what identity box will they fit? Shall we not judge them by the content of their character, not by their skin colour?

Whatever happened to Martin Luther King's dream of freedom? "Free at last. Free at Last!"

In the same year that apartheid was devised to keep captive the rainbow nation; in the same year that Israel became one of the nations, free at last from the horror of the holocaust where six million Jews were executed and millions more enslaved, in that same year, a popular musical hit Broadway for the first time. Its premise was that "race" is something we made up to suit a false, misleading theology and worldview. To explain such a monstrous lie, it can only be progressed by teaching it to our children; at home, schools and in universities.

    You've got to be taught to hate and fear,

    You've got to be taught from year to year,

    It's got to be drummed in your dear little ear

You've got to be carefully taught!

You've got to be taught to be afraid

Of people whose eyes are oddly made,

And people whose skin is a different shade — You've got to be carefully taught.

You've got to be taught before it's too late,

Before you are six or seven or eight,

To hate all the people your relatives hate — You've got to be carefully taught!

Rogers & Hammerstein. South Pacific. 1948

It would be good to end this story there, but there is one more speech, his greatest speech, to consider, one more twist to relate.

The "I Have a Dream" speech was in August 1963. Many things happened after that, many good things. Progress was made.

Nearly five years later, Martin Luther King made what many consider to be his greatest speech.

He started by asking a question – if God could place me anywhere in human history, where would I wish to be?

He asked would it be in Egypt, watching God free the Hebrew slaves. But no, he couldn't stop there.

He asked, would it be in Greece to watch the great orators tangle with the truths of humanity?

He asked would it be in the great days of the Roman Empire, and great government, but no, he couldn't stop there.

He asked, would it be to watch as Martin Luther nailed his thesis to the door of the traditional church? But no, he couldn't stop there.

He asked if it would be possible to watch President Lincoln sign the Emancipation Project in 1863. But no, he couldn't stop there.

And finally, his Spirit settled in the current day.

Cometh the man cometh the hour.

"Strangely enough," he said, "I would turn to the Almighty and say, if you allow me to live just a few years in the second half of the 20th century, I will be happy."

I mentioned earlier I recently had a near-death experience. Many, in fact, I had multiple cardiac arrests in one night. (The chances of recovering from more than one is remote.) The following day, in hospital, having survived the ordeal, I was also diagnosed with lung cancer. I was evacuated by Air Ambulance to Melbourne, 600 kilometres away, where, in time, I recovered. (Read the full magazine article.)

For many months, I struggled with the "why?" Yet what I experienced that night has left me with an overwhelming sense of peace.

Martin Luther King finished his greatest speech:

> Well, I don't know what will happen now. We've got some difficult days ahead. But it really doesn't matter with me now, because I've been to the mountaintop.
>
> And I don't mind.
>
> Like anybody, I would like to live a long life. Longevity has its place. But I'm not concerned about that now. I just want to do

God's will. And He's allowed me to go up to the mountain. And I've looked over. And I've seen the Promised Land. I may not get there with you. But I want you to know tonight, that we, as a people, will get to the promised land!

Martin Luther King. Memphis, Tennessee. April 1968.

The next day, King would be assassinated by a bigoted jail escapee, James Earl Ray. Ray fled the scene and months later was arrested at London's Heathrow airport, where he planned to migrate to what was then the racially segregated Rhodesia. He would have felt at home there.

I would like to think we are better than that. I believe we can get to the Promised Land.

Summary:

These have been the days of my life, so far. I have witnessed health and happiness, death and destruction, tragedy and terror, in those days. Much of that has been imposed by humans upon humans. This can be changed, one person at a time, one day at a time. It can start with you and I. In the final part of this book, we are going to explore the good things, the better possibilities. Why are we here? A better story.

PART 3

Why Am I Here?

The two most important days in life
are the day you were born
and the day you find out why.
Mark Twain

CHAPTER 19

The Pale Blue Dot.

In 1977, NASA launched Voyager 1 on a majestic trip through the Solar System and beyond. Some 13 years later, as a parting gift, it took one of the most striking images ever and sent them back to The Luckiest Planet. The following is from a NASA article: What Is the Pale Blue Dot?

> Voyager 1 was speeding out of the solar system — beyond Neptune and about 3.7 billion miles (6 billion kilometres) from the Sun — when mission managers commanded it to look back toward home for a final time. It snapped 60 images to create the first "family portrait" of our solar system.
>
> The picture that would become known as the Pale Blue Dot shows Earth within a scattered ray of sunlight. Voyager 1 was so far away that — from its vantage point — Earth was just a point of light about a pixel in size.
>
> Carl Sagan played a leading role in the U.S. space program. The prominent planetary scientist was a consultant and adviser to NASA beginning in the 1950s. He briefed the Apollo astronauts before their flights to the Moon…. Sagan also was a member of the Voyager Imaging Team. He had the original idea in 1981 to use the cameras on one of the two Voyager spacecraft to image Earth. He realized the images might not show much because the spacecraft was so far away. This was precisely why Sagan and other members of the Voyager team

felt the images were needed — they wanted humanity to see Earth's vulnerability and that our home world is just a tiny, fragile speck in the cosmic ocean.

On Feb. 13, 1990, Voyager 1 warmed up its cameras for three hours. Then, the spacecraft's science platform was pointed at Neptune, and the observations began.

After Neptune, it took images of Uranus, Saturn, Mars, the Sun, Jupiter, Earth, and Venus. The Earth images were taken at 04:48 GMT on Feb. 14, 1990, just 34 minutes before Voyager 1 powered off its cameras forever.

It took until May 1, 1990 — and four separate communications passes with NASA's Deep Space Network — for all the image data to finally arrive back on Earth. Voyager 1 had captured images of six of the seven planets targeted and the Sun.

In his subsequent book "Pale Blue Dot: A Vision of the Human Future in Space," Sagan wrote of the image: "That's here. That's home. That's us. On it, everyone you love, everyone you know, everyone you ever heard of, every human being who ever was, lived out their lives… on a mote of dust suspended in a sunbeam."

"There is perhaps no better demonstration of the folly of human conceits than this distant image of our tiny world," he said. "To me, it underscores our responsibility to deal more kindly with one another and to preserve and cherish the pale blue dot, the only home we've ever known."

So far, we have discovered where we come from and how we came to be here.

We have learned how humans have inflicted suffering upon fellow humans through slavery, war and discrimination against those "made in God's image."

Now, we can turn our attention to Planet Earth, the most unique home in the universe. There is no other place in all of Creation quite like Planet Earth.

We humans have been on this luckiest of all planets for millennia. So, why are we here? What is our purpose, our mission, and our mandate? How should we conduct ourselves and manage the Earth, our home?

Some think those are difficult, complicated questions. Not so!

The planet is not a "living thing," as some who worship it under the guise of different identities, like Gaia, claim. The Earth was commanded by God to create an environment conducive to life – and what a wonderful planet it has produced. However, the complexity of systems within systems is difficult even for modern science to explain. The Earth is made of inanimate minerals and surrounded by a life-supporting atmosphere. The systems operating inside this structure are called climate; the variations experienced from region to region are caused by weather. And sometimes, the Earth can produce devastation. We shall discover that this planet is dynamic, dangerous, and naturally deficient in ways that demand we struggle to survive. To find out why we are here, we must return to the book of Genesis, the Book of Beginnings!

* * *

Moses recorded the story of Creation 3500 years ago – as true and accurate today as it was then. But that was only the beginning – here is the rest of the story as portrayed in Genesis 1;27-31 CJB

> So God created humankind in his own image;
> in the image of God he created him:
> male and female he created them.

God blessed them: God said to them, "Be fruitful, multiply, fill the earth and subdue it. Rule over the fish in the sea, the birds in the air and every living creature that crawls on the earth." Then God said, "Here! Throughout the whole earth I am giving you as food every seed-bearing plant and every tree with seed-bearing fruit. And to every wild animal, bird in the air and creature crawling on the earth, in which there is a living soul, I am giving as food every kind of green plant." And that is how it was. God saw everything that he had made, and indeed it was very good. So there was evening, and there was morning, a sixth day.

And on the seventh day, God rested, forever defining the seven-day working week we observe today. The Jews still keep Holy the seventh day, as they were commanded. But that was far into the future: here, at the beginning, God is dealing with all of mankind.

He was pleased; he looked at His work and declared it very good. He had a personal relationship with these two made-in-his-image spiritual-earthy beings. But it did not last - we all know how Adam and Eve ate the forbidden fruit and were cast out of the Garden, the paradise created just for them. Suddenly, their world changed from a blissful life to a life of strife, struggle and toil. From then on, things went downhill for the first humans, from whom we are descended. As one of our Australian Prime Ministers said, "Life wasn't meant to be easy!"

But God still blessed them and their descendants, explaining their new responsibilities. They were given the freedom to have many offspring, increase and multiply. To have dominion over everything; the animals, insects, birds, fish and reptiles. To take over the management of this planet and every living thing on it. Their mandate was to manage the whole planet, to have a considerable impact, to be gardeners, not caretakers, as some would have it. To eat fruit, grow crops from seed, farm animals, and cultivate the land. From the beginning, all these things required an impact on the planet. Mankind is here to change things for their own good. God told humanity to flourish and prosper in the abundance they created with their own labour in their new home. And so it is.

To the woman he said:

> "I will make your pains in childbearing very severe;
> with painful Labour you will give birth to children.
> Your desire will be for your husband,
> and he will rule over you."

To Adam, he said, "Because you listened to your wife and ate fruit from the tree about which I commanded you, 'You must not eat from it,'

> "Cursed is the ground because of you;
> through painful toil you will eat food from it
> all the days of your life.
> It will produce thorns and thistles for you,
> and you will eat the plants of the field.
> By the sweat of your brow

> you will eat your food
>
> until you return to the ground,
>
> since from it you were taken;
>
> for dust you are
>
> and to dust you will return."

Adam named his wife Eve, because she would become the mother of all the living.

Genesis 3:16-20 New International Version

And so it was. Eve gave birth to two boys, Cain and Abel. All went well until they grew to manhood. Cain was a man of fields and crops; Abel was a shepherd of flocks. And they came with sacrifices to the Lord. Cain with grain from the fields, Abel with a firstborn from the herd. But because God favoured the sacrifice of Abel, there was a dispute – anger prevailed, and Cain murdered Abel. God sent Cain from his Presence to be a wanderer forever. And so it was.

Do I hear you say, "Well, what about the flood? Did that not get rid of this Covenant between God and all mankind? Not so. After the flood, this happened. From Genesis 9:1-7 Complete Jewish Bible.

> God blessed Noach and his sons and said to them, "Be fruitful, multiply and fill the earth. The fear and dread of you will be upon every wild animal, every bird in the air, every creature populating the ground, and all the fish in the sea; they have been handed over to you. Every moving thing that lives will be food for you, just as I gave you green plants before, so now I give you everything — only flesh with its life, which is its blood, you are not to eat. I will certainly demand an accounting for the blood of your lives: I will demand it from every animal and from

every human being. I will demand from every human being an accounting for the life of his fellow human being. Whoever sheds human blood, by a human being, will his own blood be shed, for God made human beings in his image. And you people, be fruitful, multiply, swarm on the earth and multiply on it."

Once again, God reminded humanity that it was his wish for them to flourish. This was His divine blessing. Yes, there were thorns and thistles in the fields, but mankind is expected to prosper in the abundance they create with their sweat and labour. His plan was for humanity to live fulfilled lives full of meaning, wellness and happiness. To literally "swarm upon the Earth and multiply upon it." But this time, because of the murder of Abel by Cain, He told them that the shedding of blood between humans was wrong and would require a reckoning. In The Creation Mandate: What Does it Mean to be Fruitful and Multiply? It says:

> God's power is infinite, so the creation mandate reaches all people through time and space. Not even the Fall overthrew His mandate to fill the earth and be stewards of creation. The long line of impressive civilizations throughout history, with all their remarkable achievements, technological advancements, and beautiful artwork, are evidence of God's power and kindness to all humanity,

***

Summary:

It is God's blessing that we be fruitful and prosper; human flourishing is God's idea! That promise and blessing have never been revoked. It is current today.

Our responsibility <u>as a race</u> is to have children and for them to have children. God's will has us populate the whole earth, and God has <u>created the world significantly to sustain us.</u> But it will take effort and sweat. It won't be easy.

Our obligation to the planet? To manage the whole planet so that humans flourish and prosper sustainably. To ensure that <u>all life</u> is properly cared for and protected. "The fear and dread of you will be upon every wild animal, every bird in the air, every creature populating the ground, and all the fish in the sea; they have been handed over to you. Every moving thing that lives will be food for you, just as I gave you green plants before, so now I give you everything." God's plan is for humans to be in charge of nature and work with nature – not the other way around!

We are collectively responsible for humanity –all made in His Image. In other words, how we govern and manage our people and treat other nations must be done with love and compassion. In God's eyes, all humans are created equal and should be treated the same way. "I will certainly demand an accounting for the blood of your lives: I will demand it from every animal and from every human being. I will demand from every human being an accounting for the life of his fellow human being."

Humans have God's blessing to manage and rule the planet carefully and wisely. He wanted Adam and Eve and their offspring to flourish and have children. He wishes the same for us today. He gave them the freedom to be innovative. In these final chapters, we will explore the concept of human flourishing and how it pertains to our lives today. What to do about the climate? What is the moral case for global, affordable energy, and why should we care? What should be our response to humans' current de-population of the planet? (That's right, the de-population bomb is one of the most serious threats to human civilisation!) How do freedom and innovation lead to a more prosperous world? And what role in our understanding of these issues do God and the Bible play?

CHAPTER 20

Here's a Real Climate Catastrophe!

I grew up in the North Island of New Zealand and often drove around or holidayed on the shores of Lake Taupo. It is one of the biggest freshwater lakes in the southern hemisphere. I have spent many happy days fly fishing with my brother, Hamish, on the iconic Tongariro River, which feeds Taupo.

While driving home around the lake after a day's fishing, he asked me if I knew how Lake Taupo was formed? He explained that Lake Taupo was formed by two of the most significant eruptions in recent history, "*Hatepe*," in the second century AD, created unbelievable levels of destruction. It spewed 120 cubic kilometres into the atmosphere and caused a colossal mega-flood. Fortunately, the country was uninhabited at the time. Yet, because it occurred in the southern hemisphere, the ash cloud, which affected the atmosphere and weather for a decade, did not drift over the equator or affect Asia, Europe, or North America. So, it could have been worse!

It was one of the most powerful eruptions on our planet in the last 5000 years. This super-volcano has a significant eruption approximately every 1000 years, so it's long overdue. But that wasn't a real climate catastrophe because it impacted so few people.

Here's a real climate catastrophe.

***

Spending five years in the attic of his suburban home in England, David Keyes, a writer on history and archaeology, immersed himself in a worldwide investigation to discover the cause of the most catastrophic climate disaster affecting humans in recorded history. David says in [536 AD: The Year the Sun Disappeared! David Keyes. History Hit Network](#).

> 1500 years ago, something terrifying happened to the world's climate, something nobody could understand. The Sun began to go dark. Rain, the colour of blood, poured from the skies; clouds of fine dust enveloped the Earth. Winter gripped the land for two years, followed by drought, famine, plague, and death. Whole cities were wiped out, civilisations crumbled, and nobody knew what had happened. It was a catastrophe, a catastrophe that affected millions and millions of people all around the world. But what was it?
>
> This mid-6th century catastrophe was the most crucial date in the history of the past 2000 years. It really did lay the foundations of the world we live in today.

David has spoken with over 80 experts on drought, famine, flood and disease, ecological disasters, epidemics and ancient wars from the 6th and 7th centuries AD from all over the world.

At a conference in 1994, he became aware of an anomaly in history in the mid-6th century. Mike Bailey was a *"dendrochronologist,"* an expert in tree rings. By studying the ring formation of very old trees, it's possible to discover details of the climate of ancient times. They indicate how trees were witnesses to the globe's climate and weather. A good year has wide rings, and a bad year has a narrow ring. He explained how all the tree rings in the world really went haywire around the middle of the sixth century. The rings around 535AD displayed a climate gone wrong. The dating is exceptionally accurate. The computer-generated data records the annual temperature yearly for the past thousand years. Data from Europe, Asia, and the Americas all showed these were the worst years in the past 5000 years.

Something big happened in the sixth century. What was it? A more detailed examination showed extreme, prolonged cold periods in the mid-6th century. Archaeological evidence from ancient sites in Ireland suggests that for that decade, it was so cold crops failed, and the population had to rely on fishing and hunting.

David Keyes was intrigued. What had caused this extreme climate? Were there written records? The dominant civilisation at that time was the Romans. Contacting academics in Constantinople, the Roman capital at the time, he found many accounts of bizarre weather around that period. He says

"There was a sign from the Sun, the like of a which had never been seen or reported before; the Sun became dark, and its darkness lasted for 18 months. Each day, it shone for about four hours; still, this light was only a feeble shadow. Everyone declared that the Sun would never recover its whole light again."

David found widespread records of frosts, darkness, and failing crops throughout Europe and Asia, China, Korea and Japan for that period. All made mention of this climactic event. " In 540, the Japanese great king wrote, "Food is the basis of the Empire. Yellow gold and ten thousand strings of cash cannot cure hunger; what avails a thousand boxes of pearls to him who is starving of cold. The ancient Nanshi chronicle of southern China records that "yellow dust rained down like snow; it could be scooped up in handfuls."

Imagine living in the middle of the 6th century and suffering a climatic catastrophe, an event so horrendous that trees hardly grew for years, and the Sun was dimmed. Whatever it was, it would have taken thousands of cubic miles of dust to be hurled into the atmosphere, causing this permanent winter. Was an asteroid, comet strike or volcano responsible?

Scientists from NASA calculated that from the amount of dust recorded in the atmosphere, it would have to be an asteroid over four kilometres across. It would take an even bigger comet to create the same effect. A comet is mostly ice and gas – it was calculated that it would take a six-kilometre-wide comet to achieve the same damage. David Keyes explains:

> When it was just over two days from impact, it would only be seen as a faint star in the night sky. As it approached us, as it got closer and closer, we'd slowly see it brighten and grow larger until about 30 minutes before impact, it would be about the brightest thing in the sky. And by then, of course, we believe everybody would have noticed it, but we wouldn't have been able to do anything about it. Now the time it takes for that asteroid to travel from the top of the Earth's atmosphere until it reaches sea level is only eight seconds, so we'd see this brilliant fireball; of course, making no sound because it's travelling about 20 times the speed of sound. The first sound we would hear would occur minutes after seeing the massive flash of light when the asteroid strikes the Earth's surface and is instantly vaporised in a ginormous fireball.

Could such an event have happened without being noticed? No civilisation at the time recorded any such event! In addition, there is no scientific evidence of what would need to be a large crater from the sixth century. But, the lack of a crater alone does not rule out a comet or asteroid strike. Seventy per cent of the Earth is covered in water. Could it be that the impact was on the oceans? If the asteroid had landed in the sea, the initial wave caused by the impact would have been miles high; there would have been humongous tidal waves that would have swamped the coasts of whatever ocean it struck. The tidal damage would have travelled miles inland. Again, no civilisation recorded such an event, and scientists haven't detected any significant interruption to the growth of coastal plants at this period. There doesn't appear to be any evidence of an asteroid or comet strike on Earth then.

David ruled out asteroids and comets! And that left volcanoes!

David contacted scientists who worked with ice cores drilled from the Arctic ice shelf. They were able to test the frozen ice for traces of sulfuric acid in that period, which was a sure sign of volcanic activity. And bingo, around 536AD, there are traces of significant volcanic activity – not just traces, but evidence of a colossal volcano erupting around that same time. But where and which volcano created this climate catastrophe? David said;

> It could only happen near the equator, as only equatorial winds can spread dust over both hemispheres. But there are over 90 equatorial volcanoes. Could David Keyes discover which one caused the mayhem of the sixth century? He began to narrow the search for the 6th-century volcano. He knew the highest concentration of large tropical volcanoes was in an arc

straddling Southeast Asia from India to Sumatra, Java, the Philippines, and Japan.

First, he turned to the greatest civilisation near this area, which was then producing written records – so he travelled to China. With great excitement, he started looking to see if there was any trace of anything happening in 535. In fact, in February 535, there was a record of a loud bang, a huge thunderous sound coming from the southwest. With this one, there was no mention of lightning or anything; it was merely a mysterious entry in which they only referred to a thunderous noise. Interestingly, that points straight towards the Indonesian area where all those volcanoes are.

For the Chinese to have bothered to record such a sound, it must have been an exceptional one-off event, but could the sound of a volcanic eruption have travelled 3000 miles from Indonesia to China? To help locate the volcano, David Keys asked experts at Los Alamos laboratory in New Mexico to explain the physics of long-range sound travel. "We know that near the volcano, the sudden explosive eruption provides a shock wave in the near field, but that propagates out going out to thousands of miles, even as far as China!"

536 AD: The Year the Sun Disappeared! David Keyes. History Hit Network

David Keyes and his team have narrowed the source of the eruption to somewhere in Indonesia. But where? Could there be any written evidence? Not much of the historical record has survived from that era. Still, he found a fascinating document in The Royal Palace in Java. A manuscript called the Book of Kings was based on ancient sources. It describes an extraordinary event around the middle of the first millennium. ".. a mighty thunder answered by a furious shaking of the Earth, pitch darkness, thunder and lightning. Then came forth a furious gale, together with a hard rain and a deadly storm darkening the entire world. In no time, there came a great flood. When the water subsided, it could be seen that the island of Java had been split in two, thus creating the island of Sumatra."

Had Keys struck gold with the Book of Kings? Geophysicists he consulted said the story accurately described a major volcanic eruption, but which of the many Indonesian volcanoes was the Book of Kings describing? There were over 90 to choose from. As a clue, the only central volcano in that specific area between the islands of Sumatra and Java is the legendary Krakatoa, the world's most notorious volcano, which last erupted in 1883. But could Keys prove Krakatoa was the culprit?

An Icelandic volcanologist, Professor Harolder Sigurdsson, was an expert in the field. He left work in the USA, flew to Indonesia and joined the chase. He had been to Krakatoa many times and knew there was a mystery eruption hidden deep in its past, far more significant than anything experienced in recent times. Krakatoa is part of a group of uninhabited rainforest islands lying west of Java and just south of the equator. It's also the scene of the most famous volcanic eruption of recent times; in 1883, Krakatoa blew itself apart, killing 36,000 people on the mainland and affecting the world's climate for months.

After months of searching, they found evidence of a super crater on the ocean floor close to Krakatoa. Was this the culprit? Did this cause years of cold and famine in the mid-sixth century? Finally, they managed to find enough samples that could be carbon-dated to close to 536AD – had they found the volcanic source of so much human death and misery?

It took David Keyes five years of study, travel, and research. Still, Krakatoa is now the most likely culprit - the super-volcano that erupted in 535 AD. It would have produced a dust cloud that enveloped the world. It would have been one of the most dangerous spectacles ever seen; a 30-mile-high column of ash and dust that brought *global climatic catastrophe*, darkness, drought, frost and famine, and ultimately chaos and war. It was a natural catastrophe that would change the course of human history. The amount of power generated by this eruption would have been equivalent to around *2,000 million* Hiroshima-sized nuclear bombs! The eruption of this ancient Krakatoa is something mankind has never witnessed before, perhaps hundreds of times larger than any volcano that's ever been seen! Scientific evidence suggests that when this massive volcano erupted in the tropics, it threw up so much ash that it turned summer to winter, crops failed for years, drought and famine gripped the land, and millions died.

But David Keyes was not satisfied. He scoured the world for records, evidence of this human calamity – and found plenty. In the following decade, he found evidence of famine on every continent, changing farming practices, and poverty that led to wars over land and borders. Whole civilisations collapsed and were replaced.

Perhaps worst of all, the first Bubonic disease ever recorded in history – began in Ethiopia. The continuing cold created a perfect environment for the virus, and The Bubonic Plague spread like wildfire, killing millions.

This was a true climate catastrophe! And it changed the course of human history forever.

Summary.

> *There is a lot of talk about climate catastrophe today. It is mostly talk, with no true understanding of the climate or what such a catastrophe would really look like.*
>
> *A real climate catastrophe could happen at any minute, without warning, and turn our planet into darkness and desolation.*
>
> *Let's be thankful for the wonderful, stable climate we enjoy today, on this luckiest of all planets!*

CHAPTER 21

For the times, they are a ' changin'!

The line it is drawn
The curse it is cast
The slow one now
Will later be fast
As the present now
Will later be past
The order is rapidly fadin'.
And the first one, now
Will later be the last
For the times they are a-changin'

Song by Bob Dylan 1964

In the present day, the debate about climate and energy has become intertwined. But in reality? They are separate issues, meeting only on the edges of empirical science. It is true – many scientists conclude that climate change threatens humanity. But many do not. But the times are changing. We say, give us the facts, then it is a sensible decision to be made by politicians and the people, with due regard to the upsides and downsides; a balanced debate. Let's address the current concerns about climate change, respect true facts and scientific predictions, and then discuss the current global energy crisis (which is really just a problem, to be solved as we have so many others.) But there can be no sensible debate when an almost cult-like religion is proposing changes and actions from a spiritual viewpoint, unable and unwilling to address the actual science, unable and unwilling to listen or debate.

The HO$_2$ Coalition is an organization of prominent concerned scientists formed to teach, educate, and explain the truth about climate-related issues. They have a most informative website. It was founded by Dr Patrick Moore, the founder of Greenpeace, who reluctantly left that organisation after many years because he disagreed with its pathway and the role of the environmental movement. William Happer, himself a concerned scientist, recently wrote this:

> The best way to think about the frenzy over climate is to consider it a modern version of the medieval Crusades. You may remember that the motto of the crusaders was "Deus vult!" (which means) "God wills it!" It is hard to pick a better virtue-signalling slogan than that. Most climate enthusiasts have not gone so far, but some claim they are doing God's work. After decades of propaganda, many Americans, perhaps including some of you here today, think there really is a climate emergency. Those who think that way, in many cases, mean very well. But they have been misled. As a scientist who actually knows a lot about climate (and I set up many of our climate research centres when I was at the Department of Energy in the early 1990s), I can assure you that there is no climate emergency. There will not be a climate emergency. Crusades have always ended badly. They have brought discredit to the

supposed righteous cause. They have brought hardship and death to multitudes.

William Happer The CO$_2$ Coalition.

Dr Ian Plimer is an Australian Geologist and author. He is a professor emeritus at the University of Melbourne and an expert on climate change. He is held in high esteem by the current Trump administration. He says:

> Very few people know geology, but geology tells us about the past. It tells us the history of the 4,567 million years we've had on Earth. And the one thing we do see is massive climate change. We've had a couple of periods when the planet was covered with ice, and we had ice at the equator and sea level, which was many kilometres thick. So that poses some questions. What drove this significant climate change? And even more so, how did we get out of the snowball earth? And so, if we're looking at modern warming, we've *got* to be able to look at what happened with past warmings and how these past warmings have been natural.
>
> So, what component of today's warming is natural, and what is caused by human activity?
>
> So, I look at the past climates. I look at the fact that the Earth has been a warm, wet greenhouse planet for over 80 per cent of time (since Creation). Carbon dioxide *has been* a major gas in the atmosphere; now, it's only a trace gas. Ice is a scarce rock and has been on planet Earth for *less than* 20 per cent of the time. And we are currently in an ice age. The best thing that can happen in an ice age is to have some natural warming. I

look at the cycles of climate, driven by where the continents are, driven by what's happening in the galaxy, driven by the Earth's orbit, driven by oceanic cycles, driven by cycles of the Sun, and we can see that within the last, say 50 million years, we've had a cooling trend.

We've had slightly warmer and slightly cooler periods within that cooling trend. Still, we've been cooling for an extended period. We've been in an ice age for the last 34 million years. During this time, we've had spikes of warming and cooling. We know this from historical times. We had the Minoan warming (about 3300 years ago.) Then we had cold times. Then we had the Roman warming (about 2200 years ago.) Then, we had the cold times of the Dark Ages. Then we had the Medieval warming (about 1000 years ago) and the cold times of the Little Ice Age, which finished in 1850, only 175 years ago!

And those who want to frighten us witless are looking at just one parameter, carbon dioxide. And they're looking at humans emitting *traces of a trace gas* into the atmosphere, which in the past had hundreds to thousands of times more carbon dioxide than now. When there was very high carbon dioxide, we experienced global cooling during that period. We had ice ages. We had snowball earth. So, what I tried to do in my chapter and what the others tried to do is to show that this is a very complex phenomenon and climate change cannot be related to traces of a trace gas, which is plant food that humans might emit. It's far more complicated than that.

Ian Plimer Interview Climate Change: The Facts 2025

Major contributors to climate change are radiation from the Sun, the Milankovitch cycles controlled by the gravitational pull from Jupiter, and volcanic eruptions, mostly undersea, that pump massive amounts of CO_2 and H_2O (water vapour) into the oceans and then into the atmosphere.

Dr Plimer says that "If we change the shape of the ocean floor, such as putting a big dirty volcano on the ocean floor, we will divert ocean currents carrying heat, and so we change the weather and ultimately change the climate. If we change the position of the continents, something we've known about for 50 years or so, we're changing how ocean water moves. It may get diverted by a continent appearing here, there, or everywhere. And this is what's happening at present with Antarctica. We do not have warm water from the equatorial areas into the polar areas. Antarctica's continent has a circumpolar current that stops warm water from coming. The continent that sits over the South Pole is glaciated. And until we break up that continent, a process which is happening, and has happened for a hundred million years, until we break up that continent, then the oceans will drive the fact that Antarctica has an ice sheet."

Rising CO_2 has never been a factor in global warming, and often, the reverse has been the effect. Carbon Dioxide is *not* a toxic gas; it's a trace element in the atmosphere. Carbon is on the Periodic Table and one of the underpinnings of life on Earth. It has been in much higher concentrations, and the Earth has flourished during those periods.

Dr Patrick Moore explains that carbon dioxide (CO_2) is a minor trace gas in the atmosphere and plant food and is essential to the growth of plants and trees. If the level of CO_2 in the atmosphere drops below 150ppm (Parts Per Million), plants will die, and the planet will become a desert. Not many people realise that before the industrial revolution, when we started burning fossil fuels, CO_2 nearly reached that tipping point.

Many scientists, including Dr Plimer and Dr Moore, agree that the planet would have been a catastrophe without the increase in CO_2 over the past 200 years. A real climate catastrophe! Carbon dioxide is necessary for photosynthesis, a plant's method of inhaling CO2 and breathing out H_2O. It is plant food – market gardeners worldwide pump it into greenhouses to increase growth. According to a recent satellite study of the globe by Australia's CSIRO, the rise in CO_2 levels (now about 420ppm) has contributed to a massive 15% increase just this century of forest and vegetation across the globe, *particularly* in the arid areas of Sub-Sahara Africa, Asia, and Australia. This substantially benefits life on Earth and far outweighs the downside of a possible couple of degrees increase in temperature. (Which will actually help most of the population globally.) The climate is a dynamic system; it's constantly changing. And as we will discover, change is improving our quality of life – and the planet!

Contrary to popular belief, climate-related disaster deaths have plummeted 98 per cent over the last century, as CO_2 levels have risen from 280 ppm (parts per million) to 420 ppm. Temperatures have increased by a mere 1° C.

As author Alex Epstein says, "And if you look into this issue, you will indeed find a huge range of conclusions among researchers about which level of warming leads to what consequences—including the conclusion of the world's leading climate economist, Nobel Prize winner William Nordhaus, that (even) 2°C is not catastrophic and that passing policies to prevent it would do far more harm than good."

Bjørn Lomborg is a Danish political scientist, author, and think tank Global Consensus Centre president. He is the former director of the Danish government's Environmental Assessment Institute. He became internationally known for his best-selling book *The Skeptical Environmentalist* (2001) He says:

> Fewer and fewer people die from climate-related natural disasters. This is even true of 2021, despite breathless climate reporting. Over the past hundred years, annual climate-related deaths have declined by more than 96%. In the 1920s, the death count from climate-related disasters was 485,000 annually. In the last entire decade, 2010-2019, the average was 18,362 dead per year, or 96.2% lower.
>
> In the first year of the new decade, 2020, the number of dead was even lower at 14,893 — 97% than the 1920s average.
>
> Bjørn Lomborg Blog Post July 26,2021.

Yes, the climate is changing, but as we always have, humanity is also adapting and changing. The planet is flourishing with higher but, nevertheless, historically low levels of CO_2. Here is just one example from Dr Patrick Moore:

> Forestry provides one of the most perfect examples of hypocritical political correctness, preaching against using the most abundant renewable resource while at the same time telling people to use more renewable resources. The same hypocritical attitude to renewables. Forests and the trees that define them are the most complex systems in the universe.

However, trees are no different from farm crops; if the demand for wood is steady and strong, private and public landowners will plant trees to supply that demand. Take note: this is the polar opposite of the contention that the way to save forests is to stop cutting.

Patrick Moore. Ph.D. Confessions of a Greenpeace Dropout.

Trees are one of the largest carbon storages globally. They are a renewable source of carbon capture, nature's way. We decrease our carbon footprint when we cut down trees and replace them. Before we can have dominion over the planet, we need to know how this most complex of all systems operates. Obviously, we must do this in a sustainable way, respecting old growth and marginalised areas.

That said, it is worth noting that even within the environmental sphere, there is plenty of positive news. Between 1982 and 2016, for example, the global tree canopy cover increased *by an area larger than Alaska and Montana combined*. As Reason magazine science correspondent Ronald Bailey put it, "Expanding woodlands suggests that humanity has begun withdrawing from the natural world, which will provide greater scope for other species to rebound and thrive.[6]" It is clear that with innovation, better practices and improved crop genetics, less land is being utilised to grow even more food. This is an excellent win-win outcome.

∗∗∗

There is a hidden agenda behind the environmentalist's "human impact" philosophy. Even if we came up with an emission-free energy source (e.g. nuclear or, in the future, fusion), they would still

---

[6] Superabundance. Marion L Tupy.

be against it. They want us humans gone. As Albert Einstein said, "Only two things are infinite: The universe and human stupidity, and I'm not sure about the former." Listening to some modern environmentalists is like listening to a doctor who is on the side of the germs. The ultimate aim is to rid the planet of humans! No joke

But God gave this planet to us. To flourish and prosper.

***

One of the most insidious and dangerous media-driven events happens when scientific facts get distorted and misquoted for headlines and clickbait. This is not just a modern phenomenon; consider this highly emotional report from 1970, 75 years ago:

> The mainstream publication Life magazine reported that, because of particles emitted into the air by burning fossil fuels, "Scientists have solid experimental and theoretical evidence to support . . . the following predictions: In a decade, urban dwellers will have to wear gas masks to survive air pollution. . . . By 1985 air pollution will have reduced the amount of sunlight reaching earth by one half." The story was the same in the realm of water quality. According to Ehrlich in 1970: "The oceans will be as dead as Lake Erie in less than a decade." And "America will be subject to water rationing by 1974."
>
> Fossil Future. Alex Epstein

These "predictions," which create chaos and panic in the general populace, are based on something called "Scientific Modelling." Various factors and data are fed into a computerised model, and the resulting catastrophic predictions are fed to media eager to please, promote and print. The problem is that this planet and the climate are so complex and have so many varying factors that so far, modelling can only predict the weather out to a few days, and indeed not the climate for decades in advance. For example;

> In 1989, the Associated Press reported a set of climate predictions from the New York office director of the UN Environmental Programme, Noel Brown, including that "entire nations could be wiped off the face of the Earth by rising sea levels if the global warming trend is not reversed by the year 2000," and that "the most conservative scientific estimate" would be that "the Earth's temperature will rise 1 to 7 degrees in the next 30 years." The temperature rise since Brown's "conservative" prediction has barely reached the low end of his "most conservative scientific estimate." And we are more than twenty years past the year 2000, still <u>with record fossil fuel use</u>, and there is nothing resembling "entire nations" being "wiped off the face of the Earth by rising sea levels."
>
> Fossil Future. Alex Epstein

Ninety-one-year-old Freeman Dyson is revered today as a father of modern climate science. A mathematical physicist, he is one of the most renowned scientists in the United States, having been around for six decades. He worked in the same building as Albert Einstein at Princeton University in his youth.

He agrees that the slight increase in CO_2 has had minimal effect on the climate but a massive effect on the greening of the globe:

> The Earth is actually growing Greener. This has been actually measured from satellites. The whole earth is growing Greener as a result of carbon dioxide in the atmosphere, so it's increasing

agricultural yields, it's increasing the forests, it's increasing all kinds of growth in the biological world, and that's more important and more certain than the effects on climate.... well, it's increasing just more or less as we expected when I started working on this 37 years ago. And at that time, when we thought the effects (of increasing CO_2) were maybe about 10%, and now it's probably more like 25%. After 35 years, it's essentially what we expected. Carbon dioxide in the atmosphere has gone up by 40% in that time, or something like that, yeah, up to about 400 parts per million at the moment or just shy of that, I think, and yes, about half of that is gone into raising vegetation. So, vegetation has increased on average by around 20%, and that's observed, and of course, it's essential. It is enormously beneficial to food production, biodiversity, the preservation of species, and everything else that's good.

Is Carbon Dioxide Making the World Greener? Freeman Dyson, Institute for Advanced Studies)

Freeman Dyson is critical of attempting to model the climate for future predictions. He speaks of his friend Yiro Manabi, a Japanese climate expert living in Princeton, who was the first to do climate modelling for increased carbon dioxide. He found there would be warming, but it would be far less than what is now fashionable. His modelling goes back to the late 60s. He says that "these climate models are excellent tools for understanding climate. Still, they're terrible tools for predicting climate, and the reason is simple. That there are models with only a few factors in them may be necessary, so you can vary one thing at a time.... that's why they're not suitable for prediction. The real world is far more complicated than the models, which is the challenge. I don't think any of these models can ever be predictive because climate is too complex, yes, and there are too many factors at work, and you cannot model everything…. it's way out of sight!"

Is climate change real?

Yes, of course.

Have mankind's activities caused a slight rise in CO_2?

Probably.

Is that a good thing?

Absolutely!

* * *

Summary

> *Life as we know it ceases to exist at 150ppm (parts per million) of CO2. Before the industrial revolution, at 240ppm, we came perilously close to that. Thanks to human impact, we now have a much more favourable climate for plant life. And warmer temperatures are improving, not degrading the lives of millions of people in the colder climates. Historically, far more people have died from cold than from heat.*

> *We can cope with these changes incrementally with the advances in human knowledge and innovation over that same period. Climate-related deaths have decreased remarkably.*

> *The climate is comprised of an amazingly complex set of variables. It is impossible to model future predictions accurately. Perhaps in the new age of AI, that will change. But not yet.*

We are still in an Ice Age. The rise in temperature (1°) and sea level rises (2cm per annum) over the past 150 years are welcome and well within the anticipated normal levels of climate variation. In the 4.5-billion-year history of this luckiest of planets, humans have seen far more impressive and catastrophic events than we are experiencing now.

The solution to climate change is not poverty – it is to adapt to the changing climate just as we have been doing for thousands of years.

There is not, nor will there be <u>in the foreseeable future</u>, a climate catastrophe. And certainly not a human-induced climate catastrophe.

Yes, we are experiencing a catastrophe. An energy and de-population catastrophe that no one is talking about. So, let's talk about that.

CHAPTER 22

It's <u>All</u> About Energy!

Here's a story from somewhere in the world. Maybe Africa, India or China. Even South East Asia. But somewhere.

The first light, sneaking cautiously through the window high on the wall of the mud brick hut, urged Nakaala to start another day. She had been up most of the night attending to Minosen, her six-month-old daughter and youngest of five. Jaboo, the only other survivor, was already awake; he still missed his two brothers and father, all taken by the coughing disease. She dressed in a rough smock and hurried to stir the fire, still stiff from her previous day's exertions, noting that this was the last of her wood. Soon, the fire was smoking promisingly, filling the rude single room and provoking coughing from the baby, who was also awake and begging for attention. Within minutes, hot cornmeal cakes were ready for them both.

They ate them hungrily, Jaboo taking a swig of water from the pail in the corner and a hat and threadbare coat that had belonged to his father. With a corncake for lunch, Jaboo left to face the misty chill of the winter morning. He would spend half an hour milking the two cows, sufficient for their needs and enough left over to barter for corn, eggs and other essentials. They had sold the two calves for much-needed cash only recently and for emergency food. The cows were still heavy with calf milk. He soon filled the two containers, steaming warm and fragrant in the dim morning light like the creamy moustache as he sneaked a swig.

He would spend the day with both cows in a distant pasture with abundant sweet grass. As he travelled, he was joined by other boys and their animals so that soon, a tiny herd raised a pall of red dust, followed by a boisterous mob of youths who thought they were in control and eager for the day's adventures. That could include many predators and challenges, not least lions.

By mid-morning, the Sun was well up, casting welcome shade around the pepper tree at her front door. Using her pride and joy, an old 20-litre kerosene can with a carry handle, she had spent the morning pumping and carting water from the well to fill the trough at the side of the hut used for washing clothes. Occasionally, neighbours would give her a bar of soap, which she treasured and used sparingly. But not today – it would be a deep soak for a few hours and a rough scrub. She also filled the two large pitchers inside, one for cooking and drinking and one for washing. The village was blessed with good potable water, and she counted her many blessings. But for now, she sat in the shade, back against the tree, with the baby suckling noisily at her breasts and realising, as always, she would probably not be satisfied. Nakaala entered that dreamy world between waking and sleeping, as did the child at her breast. At the height of summer, she would have to collect wood and dung for the fire *before* the heat of the midday sun made it impossible to walk and carry long distances. Mother and child slept.

An hour later, in the timber country far from the village, and with Minosen tied firmly to her back, she joined a few other women foraging for firewood and pats of dried dung left by cattle and other wild animals. The wood was favoured because it did not give as much smoke, but everything flammable was welcome. The women made a pile of timber and dung, each to their own. After a couple of hours of work, the piles were considered sufficient, and the women sat down in a circle of shade, exhausted but looking forward to an hour of rest and gossip. Nakaala was highly regarded in the village, and as she was recently widowed, she was given as much help and assistance as possible. She knew that the men were butchering a beast today and that when she returned, there would be a good hunk of fresh red meat in the hut – their ration for the week. She was a good cook, and together with lentils from a sack in the corner and herbs that she had gathered during the day, they would feast tonight.

Time was up – each woman helped the other place the bundles of wood onto their heads. As the Sun retreated, causing long shadows, they returned to the village, closely followed by the boys and their herd of cattle. The singing women in a stately, swaying single-file, each with a large bundle perched precariously on her head, often with a child strapped to her back and a toddler at foot. Nakaala and the baby, crying now continuously, entered the hut, lit the fire, and started cooking. The thick smoke billowed outside like a morning mist from high on the wall. Her last duty tonight would be to snuff out the candle before bed, itself a potent contributor to the toxic atmosphere in the hut.

Another day is done.

***

Twenty years have passed. Nakaala still lives in her mud-brick hut. She still grieves the loss of her two sons, her husband, and, more recently, Jaboo, who died of Tuberculosis in his teens. She is the only woman of her generation to have survived the smoke. In fact, she is the village elder, along with a couple of older men who survived into their forties.

But there have been significant changes.

It is late at night. Minosen and her husband Darak are in their own bedroom with two young children. Alone at last, laying her book down on her lap, she looks around, as she does most nights, seeing ghosts, seeing what once was. "But never again," she smiles to herself. Back in those days, she went to bed with the sun. Her mind drifts to *what* she thought of as "Power Day!" The day that modernity came to town, in the form of energy down a wire. Electricity. Fossil fuelled power from the new power station is only sixty miles away. It changed everything.

The twin LED globes in the ceiling lit the whole room – her eyes hovered over the walls, where they had spent years scrubbing off the remains of decades of smoke. Darak had fixed that – after they had built the two new rooms, mud brick, of course, he bought a can of white paint for the whole house and painted it white. It looked like a palace.

On the far wall was her pride and joy; next to the new sink was an electric oven with two hotplates. She cleaned it daily, rejoicing in the purity of the air and the quality and quantity of new meals she cooked. An international aid organisation had adopted the village; they said their job was to empower people, not just provide things. Aged ten, Minosen had gone to the village school, a massive building with year-round air-conditioning, and excelled. Minosen had taught her mother to read, not well, but cautiously and still progressing. The recipe book on her lap was her pride and joy, and her meals were famous throughout the village. The time she once spent foraging and carting for the fire was now spent at the school, helping her daughter, one of two trainee teachers. Every kid in town went to that school. They had computers, laptops, and something called "Starlink," allowing them to explore the outside world freely.

The three "elders" and a couple of younger men managed the town; the rest worked. They were an industrious lot – no need now for hunting and foraging. They were empowered to grow their own food. The aid organisation put an electric pump on the village well. It used village labour to supply each house with running water. They had helped dig septic tanks and cut disease and sickness in half. Then, they drilled another bore with a solar pump and an overhead storage tank and bought two men from Israel to teach them drip irrigation. That changed everything!

The cattle were now in a large fenced paddock with water, forage, shade, and access to irrigated lucerne and sorghum for fodder. No longer needing to hunt for food, the men were all gainfully employed growing and harvesting pulses, chickpeas and lentils. A small crop of wheat is used each summer for poultry. A few men went to the night school once a week with a visiting teacher from the city, revelling in the knowledge that their children would be better educated than they were.

There was a small medical centre. It is run by a qualified nurse and midwife who had moved to the village, three young women trainees, and an electrically-powered infant incubator. Quite a few children were running around the school playground who would have died in years gone by – because there was no power.

Nakaala laid her head back, closed her eyes, and drifted off to sleep, her book still on her lap. She dreamt of the old days. Empowerment! If only they could see her now?

Soon, she was softly snoring and gently smiling at the same time.

***

The argument goes like this: "The earth's climate is heating exponentially, and oceans are rising catastrophically. Therefore, at all costs, fossil fuels must be banned and replaced by renewables like wind and solar." Does that sound familiar? Is that scenario supported by facts? Are you ready to deny Nakaala and her family the dignity of empowerment? The same dignity we have had for 150 years, powered almost exclusively by fossil fuels?

> Up to 2.1 billion people still use polluting fuels and technologies for cooking, mainly in Sub-Saharan Africa and Asia. The traditional use of biomass also means households spend up to 40 hours a week gathering firewood and cooking, which makes it difficult for women to pursue employment or participate in local decision-making bodies and for children to go to school. Household air pollution created by using polluting fuels and technologies for cooking results in 3.2 million premature deaths every year.
>
> World Health Organisation

Alex Epstein is a philosopher and author specialising in energy and fossil fuels. In his latest book Fossil Future he says:

> Here's a scary fact that we seldom hear: Over 3 billion people use almost no energy, including electricity… Observe that our knowledge system, as well as our culture, has infinitely more

interest in the fate of polar bear habitats than it has in three billion people without energy.

How have we gotten so good at protecting ourselves from climate? In large part by using fossil-fuelled machine labour. We use fossil-fuelled construction machines to build sturdy buildings. We use fossil-fuelled heating machines to produce warmth when it's cold and fossil-fuelled cooling machines to create cool air when it's hot. We use fossil-fuelled irrigation machines to alleviate drought. To put the relationship between fossil fuels and our safety from the climate in a sentence: ultra-cost-effective fossil fuel energy powers the machines that produce unprecedented protection from climate.

Ah, well, you say. That was before we found out that fossil fuels were destroying the planet! Really? A 1°C rise in temperature in 150 years is a catastrophe? Where I live, the temperature rises more than that while we enjoy lunch!

And what did we learn from Genesis? What did the Lord promise Noach?

> "I will certainly demand an accounting for the blood of your lives: I will demand it from every animal and from every human being. I will demand from every human being an accounting for the life of his fellow human being."
> Genesis 9:5-6

Are we really willing to ignore God's warning, to accept this amazing lifestyle of comparative comfort and prosperity whilst leaving 3 billion people behind to live in abject poverty? Well, it's not our problem, you say? That's not what Noach heard! God made us responsible for each other.

Modernity has moved so fast that my generation, the post-war baby boomers, are the last to remember and have experienced the old ways. For instance, my wife's family were graziers in western NSW for over 150 years. Their early experiences and the settler's relationship with the local Aboriginals formed the basis of my historical novel WALKING AMONG THE STARS.

In 1974, I was the manager of the historic Marfield Station between Ivanhoe and Wilcannia. It was just Aileen, me, and my baby daughter Michelle; we were isolated and alone. My wife has cooked on a wood-fired stove, endured the outback heat over a hot oven, finished the housework and washing without any modern appliances, tended to sick kids without medical help, carted water for the household, mustered stock on foot and horseback and worked in woolsheds for weeks on end. We endured ferocious sandstorms, days on end without power, being cut off by floods for weeks, and experienced devastating drought. I have milked cows daily, mustered by horseback and sulky (a horse-drawn buggy,) cut rails and posts with axe and adze, and repaired the vehicle's yards and windmills. Each week, I would kill and butcher mutton, beef, or pork for the table and find meat for the dogs. I would cut, cart, and saw wood, then chop and split it for the stove and heating. And many other chores would be unthinkable today without modern machines and appliances.

It is sad to realise we are the last generation with those memories and of those generations before us. So, perhaps it is hard for the modern world to understand the harsh reality of stories like Nakaala and her family? But we must try. Those days lie only a few generations in the past for all of us.

***

What is Energy?

> We could take the scientific explanation, though I suspect that does not help. $E=MC^2$ was Einstein's way of explaining the phenomena of energy. The following is a quote from the NOVA Science Trust:

> "Energy equals mass times the speed of light squared." On the most basic level, the equation says that energy and mass (matter) are interchangeable; they are different forms of the same thing. Under the right conditions, energy can become mass, and vice versa."

But as I wrote in my introduction, it will take more than science to tell this story. I said that:

> "… understanding the world requires more than scientists and theologians. It needs philosophers, poets, painters, and people who dream and envision."

So, let's take a more philosophical look at the problem. What is Energy?

In the same way that carbon dioxide is "plant food," so try to think of energy as "machine food." But what is a machine? For this discussion, a machine is "an apparatus requiring energy that transmits force for a beneficial purpose." We think of machines as complex and intricate; it need not be so.

The first machine was when a man picked up a rock or a stick and used it to batter something else. Maybe another rock, perhaps an animal or maybe to open a seed or nut. Later, he used one rock to splinter another to make a spear or arrowhead. Both are machines. When fitted to a bow, a machine used for killing and hunting, it became complex, requiring more energy. Get the picture? Alex Epstein says: "A human being can kill more edible animals with a spear than he can with his bare hands using the same number of calories. A human being with a knife can gather more berries than a human without one. A human being with a container can acquire more water than one without one. Manual tools are amazing. Thus, there is no getting around that when humans must produce largely using their weak, physical bodies, the typical human being must spend enormous amounts of time to survive at a crude, insecure level. Not only is this miserable in and of itself, but it also inhibits the very thing that could improve life: time for productive innovation."

Energy is "machine food." For thousands of years, machines were basic, mostly timber or stone. But then came the industrial revolution. In Britain in the 18th century, we found a new source of energy; coal, followed by oil and gas. This transformed the world and our lives forever. Coal and oil, in fact, are all fossil fuels that store organic energy, liquid sunshine created millions of years ago by either plant or animal decay. All fossil fuels are solar energy coming from the Sun. This is the energy that drives the world. Even after decades of so-called "climate action" to "save the planet," fossil fuel usage has increased steadily yearly. Today, it supplies 87% of our energy globally. Look around you; think of all the machines you use daily. Cooking appliances, vacuums for the floors, fans and air conditioners to keep warm in winter and cool in summer. Your phones and computers. The car you drive, the buses and trains you take. The trucks that carry your goods and produce across the nation. The great ships and aeroplanes that carry passengers and trade goods to and from our shores. What about the pipelines, pumps, and power lines spanning the country that supply our water and gas and dispose of our waste? All machines; all driven by energy.

Imagine a graph of fossil fuel usage. It started at the bottom left in the 18th century, then rose exponentially to today – top right. Then, impose other factors – life expectancy in years. It follows the same curve, from the age of low 30s to the high 70s today. Is that not a miracle? Food production – same upward curve. Massive growth in the world's population – same story. Follow the graph of prosperity that reaches for the sky as billions of people escape poverty for a better life. All this is because of cheap, reliable energy and the fossil fuels that provide it. As Alex Epstein puts it:

"I will make the case that more fossil fuel use will improve the world. In this place, billions more people will have the opportunity to flourish, including to pull themselves out of poverty, to have a chance to pursue their dreams, and—this will likely seem craziest of all—to experience higher environmental quality and less danger from climate…. One of the key benefits of more fossil fuel use, I will argue, will be powering our enormous and growing ability to master climate danger, whether natural or man-made—an ability that has made the average person on Earth 50 times less likely to die from a climate-related disaster than they were in the 1°C colder world of one hundred years ago. Alex Epstein says that "even if fossil fuel elimination policies aren't fully implemented—which they won't be, given the expressed intent of China, Russia, and India to increase their fossil fuel use—<u>even widespread restrictions on fossil</u>

fuel use that fall far short of elimination will shorten and inflict misery on billions of lives, especially in the poorest parts of the world.... And if you look into this issue, you will indeed find a huge range of conclusions among researchers about which level of warming leads to what consequences—including the conclusion of the world's leading climate economist, Nobel Prize winner William Nordhaus, that a 2°C increase is not catastrophic and that passing policies to prevent it would do far more harm than good.... On a human flourishing standard, we want to avoid not "climate change" but "climate danger"—and we want to increase "climate liveability" by adapting to and mastering climate, not simply refraining from impacting climate.... These facts are: Fossil fuels are a uniquely cost-effective source of energy. Cost-effective energy is essential to human flourishing. Billions of people are suffering and dying for lack of cost-effective energy."

Our goal should not be "saving the planet" but to "improve the environment" for human beings. That's what the Bible says we should be doing. We can have the best of both worlds, that which exists naturally and the best of what we make ourselves. Humanity has made huge strides in the last two centuries, learning to protect ourselves from climate change (which is accurate but not catastrophic). With a 98% drop in climate-related deaths, we have done a great job of it! Literally millions of people are alive today because of the enormous strides we have made in "climate safety." With modern science, machines, and probably AI, we can significantly improve our "climate safety" until the next best thing comes along. And it will come along – it wasn't long ago we were killing whales to use their oil to fuel the lights in our streets and homes. We have come a long way since then. As Alex Epstein says,

"How have we gotten so good at protecting ourselves from climate? In large part by using fossil-fuelled machine labour."

***

One of our most significant dangers is from some in the "environmental movement." I am neither pro nor anti-development – I am a conservationist at heart. (More about this later) But some are "anti-human" at heart, fuelled by a self-righteous plague that thrives on spreading the lie that human impact on the planet is evil and has to be stopped at all costs. Roughly translated, "sustainable development" means "development without impact," a contradiction in terms. According to the Bible, genuine sustainable development is where the aim is for human flourishing. Not "to hell with everything else" but *including* everything else. Human flourishing is actually good for the planet. Genesis 9:2-3 CJB says:

The fear and dread of you will be upon every wild animal, every bird in the air, every creature populating the ground, and all the fish in the sea; they have been handed over to you. Every moving thing that lives will be food for you; just as I gave you green plants before, so now I give you everything.

Human flourishing is when we manage the planet for our own benefit and for the benefit of all. Across the world, humans have used the benefits of modernity, using more time created by machines to heal the planet. There are exceptions, but there has been a genuine and fruitful movement to respect and mend our wrongdoings. Look at the greening of the planet, the halting of erosion, and the responsible management of the oceans – mankind is having an increasingly positive impact. Have you noticed the decrease in smog in our major cities? Over the last twenty years, the internal combustion engine improvements have dramatically reduced air pollution. (By the way, CO_2 is a naturally occurring invisible trace gas in the atmosphere– it <u>cannot be seen and is not pollution</u>.)

* * *

What about the myth that "natural" substances are good and "man-made substances" are bad? It's simply not true - the pollution from oil, gas and coal as a fuel source is largely being mitigated. And it's the "natural" part of this organic, plant-based fuel such as sulphur, nitrogen and heavy metals; it is simply untrue that "natural" substances are safe and "man-made" substances are unsafe. When I was young, a "beach bum" in New Zealand, I had a jellyfish sting on my foot. It never really healed and was not an issue until, in my seventies, a huge cancer erupted from the wound, requiring a significant operation. This world of thorns and thistles is dangerous, not some idyllic blissful nirvana. Arsenic and cyanide are natural substances. While many naturally growing plants are poisonous. Alex Epstein says, "In reality, every phenomenon in nature, including every type of substance, is safe or beneficial below or above a certain threshold dosage and dangerous outside that range. Paracelsus said over five hundred years ago, "All things are poison and nothing [is] without poison; only the dosage determines that something is not a poison."

I was halfway through writing this book when I was flown to Melbourne and hospitalised for six weeks, including a week in the Intensive Care Unit. To pass the time, I did some research on plastics. I asked the staff, nurses and doctors, "What proportion of your equipment comes from fossil fuels?" (They use a lot, and it's mostly discarded.) Most of their guesses were very low. They were astonished and some quite horrified to learn the truth – that fossil fuels contributed to 89% of their daily equipment. I had 269 bags of antibiotics over that period, every four hours for six weeks. All those plastic bags and tubes were unusable, so they were recycled. I had a staph infection attacking my heart – I nearly died multiple times. The contents of those bags and tubes saved me and millions like me. Before we condemn something, we need to look at both sides of the proposition. Alex Epstein writes Fossil Future.

> Today's chemical engineers can "crack"—break down—the hydrocarbon molecules of oil into small parts and then reassemble them into an unbelievable variety of "polymers," including modern plastics. While you think of the "oil" in your car as being in the petrol tank, there is more oil in the car's materials than in the petrol tank. The rubber tyres are made of oil, the paint and waterproofing are made of oil, the plastic dent-resistant bumper is made of oil, the stuffing inside the seats is made of oil, and in most cars, the entire interior is one oil-based fabric or synthetic material after another—because oil is an amazingly cost-effective material to make things with.

***

So, what does the future hold? In Australia, while the cost of power soars due to the unprecedented level of wind and solar being forced into the system, our coal stations are being shut down at an alarming rate, ignoring the golden rule of cheap, affordable energy. *"Don't shut something down unless you have a replacement ready."*

Our beautiful hill country, fertile pastureland and coastline are being covered with ugly wind farms, solar farms, and high-tension power lines that straddle the country like some giant colossus, stretching as far as the eye can see. And all for what? And all while China, India and others are busily building more and more coal-fired and nuclear power stations to empower millions of their people and release them from poverty, we shout slogans at them for not "saving the planet." It is pretty apparent those countries have come to the same understanding as a growing number in the West; we are *"cutting off our nose to spite our face."*

***

Coal, oil and gas will be with us for a very long time. They are needed for 24/7 base-load power. Firming for when the Sun doesn't shine, and the wind doesn't blow. The idea that renewable energy is cheap is a myth; no country that has heavily invested in renewable energy also has cheap energy. The two are polar opposites. Countries like Australia are impoverishing themselves, driving up the cost of electricity and, therefore the cost of living. They are also driving heavy manufacturing from our shore, such as steel, aluminium, and fertilizers, to foreign countries to make the same number of products from the same fossil-fuelled power. This is insanity – these industries are the backbone of our manufacturing and essential to our security and standard of living. We are also pumping billions of dollars into "green hydrogen", a pipedream for another day.

There are over 2200 active coal-fired power plants worldwide, and many more are being built. China has the most significant number of coal-fired power stations of any country or territory globally. As of July 2024, 1,161 operational coal power plants were on the Chinese Mainland. This was more than four times the number of such power stations in India, which ranked second. China accounts for over 50 per cent of total global coal electricity generation.[7] And they are building more.

Since the beginning of the 21st century, coal has been the fastest-growing global energy source, providing about 40 per cent of the world's electricity needs. Coal is the second primary energy source in the world, after oil, and the first source of electricity generation.[8]

* * *

This brings us to the elephant in the room – nuclear-powered electricity.

I have long ceased to wonder at the stupidity of the environmental movement's opposition to nuclear. In Australia alone, we have enough uranium to last thousands of years. It is the only "non-solar" energy source, apart from fusion[9], which is the apex of power generation but is some decades off. Nuclear has zero carbon emissions and next to zero air pollution – it produces H_2O, water vapour. About 65 reactors are under construction across the world. About 90 further reactors are planned. Most reactors under construction or planned are in Asia... Over the past 20 years, 106 reactors were retired, but 102 have started operations. Today, about 440 nuclear power reactors operate in 31 countries, including Taiwan, which provide about 9% of the world's energy and growing fast.[10]

---

[7] Statista Statistics

[8] World Bank Databank

[9] Fusion occurs when two nuclei combine to form a new nucleus. This process occurs in our Sun and other stars. Creating conditions for fusion on Earth involves generating and sustaining a plasma. Plasmas are gases that are so hot that electrons are freed from atomic nuclei.

[10] World Nuclear Association.

Nuclear plants have a lifetime of 60 to 80 years, compared to 15-25 years for windmills and solar farms. So, replacement capital is required four to five times more for renewables. The big question is, what do you do with this stuff when it's finished? Just think about the enormity of that problem. All those minerals, metals, rare earths, fibreglass, steel and aluminium. What to do?

A nuclear plant footprint is also tiny compared to the millions of acres needed for solar, wind and the tens of thousands of kilometres of transmission lines. Of the forty countries that use or build nuclear plants, waste disposal is no longer a problem or is only a minor problem compared to the disposal of millions of solar panels and giant windmills. In Confessions of a Greenpeace Dropout Patrick Moore PHD says;

> "Nuclear energy is unique because it is a significant energy source not based on solar energy. Uranium is one of the rarest elements in the Earth's crust. Still, because it contains so much energy, it has the potential to provide fuel for thousands of years. One kilogram of natural uranium has the same energy as 10,000 kilograms of coal. One kilogram of uranium-235 has the same energy as 1,500,000 kilograms of coal.

Australia currently has a 65-year-old nuclear power plant in the suburbs of Sydney. I recently benefited from some of the atomic medicine it produces, saving thousands of lives annually. We are also acquiring a fleet of nuclear submarines as part of our international AUKUS treaty.

Only 4 kilograms of uranium are required to power nuclear submarines for 30 years without refuelling. The reactors used in nuclear submarines are based on highly enriched uranium (HEU), which contains 90% or more uranium-235. This enriched uranium can produce a large amount of energy in small quantities. 4-5 kilograms of enriched uranium can power a nuclear submarine for several decades, as atomic fission reactions produce tremendous energy.

So, suppose it's good enough to have a decades-old nuclear plant in our suburbs and to have our sailors sleeping next to a nuclear reactor for thirty years. Why not 24/7 atomic power stations that will address the nation's increasingly expensive power bill? The answer is that this country's opposition to nuclear power is ideological, not factual or logical. Their numbers don't add up. Instead, they point to a looming energy catastrophe! In a recent article in The Australian, What Climate Spending Really Costs the World. Bjorn Lomborg says,

> Two major scientific estimates of the total global cost of climate change have been published recently. These are not individual studies, which can vary (with the costliest studies getting copious press coverage). Instead, they are meta-studies based on the entirety of the peer-reviewed literature. One is authored by one of the most cited climate economists, Richard Tol; the other is by the only climate economist to win the Nobel Prize, William Nordhaus. The studies suggest that (even) a 3°C temperature increase by the end of the century, slightly pessimistic based on current trends – will have a global cost equivalent to between 1.9 per cent and 3.1 per cent of global GDP. To put this into context, the UN estimates that by the end of the century, the average person will be 450 per cent (more) as rich as he or she is today. Because of climate change, it will feel like "only" 435-440 per cent as rich as today.

If the climate catastrophe is greatly exaggerated, then globally, we can afford to invest much of that proposed capital to ensure those 3 billion humans are no longer the power-impoverished poor of the world.

* * *

Summary

> *Carbon Dioxide is a trace element in the atmosphere. It is plant food – necessary for flourishing plant life. When plants grow, humans flourish.*
>
> *Only 150 years ago, the atmosphere came perilously close to reaching the CO_2 tipping point for life on Earth of 150 ppm (parts per million.) Thanks to the industrial revolution, we are now at a historical low of 420ppm. And the luckiest planet is thriving!*
>
> *There is no climate catastrophe. We do not need to reduce carbon emissions to net zero. That is not supported by data or many eminent scientists. In fact, we do not need to reduce carbon emissions at all! But, if you do insist on zero carbon emissions, here is Nuclear! Perfect*
>
> *Beware the environmentalists, with their feigned moral superiority, whose innermost desire is to diminish or delete humans from the planet.*
>
> *Fossil-fuelled machines and technology have reduced climate-related deaths by 98% and will continue to reduce our "climate danger."*

Some 3 billion people are without life-changing electricity. Billions are being denied the life-empowering, fossil-fuelled electricity that we enjoy because of an "addiction" to unnecessary expensive climate mitigation ideas and policies.

Together, we can change this.

Put simply: cheap, affordable energy drives wealth and prosperity. And the opposite is true.

CHAPTER 23

The Abundant Planet

Psalm 115:16 says:

The highest heavens belong to the Lord,

but the earth he has given to mankind.

Back in the days when newspapers reigned, there was a saying, "If it bleeds, it leads!" Meaning that the most prominent headlines are reserved for grief and disaster. Have you ever wondered why we are so dominated by "bad" news? It seems part of our makeup – that the negative precedes the positive. Perhaps it throws back to our hunter-gatherer days. Imagine a hunter, fully armed with a spear and club, moving stealthily through the jungle. He hears a rustling of leaves nearby; something alive, close at hand. Rabbit or lion? What to do? Fight or flight? The adrenalin kicks in. Does he take the positive option, which is probably nothing to be worried about? Or, to be safe, the opposing view? He immediately takes flight! And, yes, that was the correct decision for someone in that perilous position to make.

Notwithstanding that bad news sells, that we are indeed wired for the negative does not make it the *most beneficial* for us. Instead, the benefits of a positive outlook are numerous; otherwise, how would humanity be able to provide wealth and prosperity for all, the result of innumerable inventions and innovations over such a comparatively short time.

In these final chapters, we will leave negativity behind and discover the remarkable story of humanity on this luckiest, most abundant Planet!

In 1970, economist Paul Ehrlich published his worldwide best seller, The Population Bomb. It was applauded by politicians, planners and the general population across the globe as a dire warning of things to come. It is still widely quoted to this day despite being fundamentally wrong. He claimed that the earth only had a finite number of resources. With record population growth, we would run out of food and resources sooner rather than later. Does that sound familiar?

> In a 1969 essay titled "Eco-Catastrophe!" Ehrlich asserted, "Most of the people who are going to die in the greatest cataclysm in the history of man have already been born.... By 1975, some experts feel that food shortages would have escalated the present level of world hunger and starvation into famines of unbelievable proportions. Other experts, more optimistic, think the ultimate food-population collision will not occur until the decade of the 1980s."
>
> Ehrlich's background in biology is relevant. In the nonhuman animal world, a sudden increase in the availability of resources leads to a population explosion, which in turn leads to the exhaustion of resources and population collapse."

Since then, for over fifty years, we have been bombarded with similar stories of woe and grief; the impending ice age, a global warming crisis, rising oceans, thousands of species becoming extinct, imminent death and famine.

Ehrlich could not have been further from the truth – yet a whole movement of enthusiastic followers and disciples sprung up and had a totally adverse effect worldwide that persists today. Where did he go wrong? As a biologist, he mistakenly equated humanity with the rest of the animal kingdom. He could not have been further from the truth regarding God's image-bearers on Planet Earth. As we shall discover, he failed to realise the unique spirit, the soul, that indwells us all, the spiritual *neshama*, providing attributes unique to humans; innovation, discovery and invention. What sets us apart from the animal kingdom are ideas and inventions that abound in the minds of those who are free to think and act for themselves. The more freedom to think, the more minds, the greater the changes.

Acclaimed author Marion Tuppy says that as the population grows in a free and open society, so does discovery and innovation, constantly revealing new resources and ways to use them. Thus, our resources are infinite on this bountiful planet, limited only by our ingenuity, thoughts and ideas. Is that not amazing? Tuppy says: "Committees don't have ideas. Algorithms don't have ideas. Machines don't have ideas—at least not yet. So far, ideas have always been a product of human intelligence. Those ideas lead to inventions, and inventions tested by the market lead to innovations that drive economic growth and higher living standards. That is why all human beings deserve dignity, respect, and liberty: to think up, experiment with, and market their ideas." He goes on to explain:

> "Superabundance is a condition where abundance increases faster than the population grows."

> Today, it's clear that what separates humans from all other species is innovation. While other animals might use a rock to crush a shell or a stick to find insects, our inventions have come

to dominate Earth, bending the latter better to conform to our desires. The vast scope of humanity's accumulated knowledge is unquestionable, but why did it take so long to produce superabundance?

For thousands of years, mankind struggled with the thorns and thistles. Yet, God gave us His Blessing and commanded us to thrive. And we are thriving – millions of people emerge from poverty yearly into a more prosperous and bountiful life.

We are currently in a time of "superabundance." According to Tuppy, an economist, personal resource abundance can be described using "Time Prices." We purchase things and services with money but pay for them with time. Time spent working. A "Time Price" is calculated by the hours and minutes of labour it takes to earn enough money to purchase an individual item. It is a more accurate way of portraying changes in wealth than just listing the nominal change in price. Therefore, it is now possible for economists to examine years of data and arrive at the same conclusion - most are now living in a super-abundant world. He explains his reasoning:

> To determine whether resource abundance is rising or falling, we analysed the time prices of 50 essential commodities between 1980 and 2020, including energy, food, materials, minerals, and metals. Our data came from three reputable organizations: the World Bank, the International Monetary Fund, and the Conference Board. We found that the average time price of the 50 essential commodities fell by 75.2 per cent. That means that the time required to earn the money to buy one unit in our basket of 50 basic commodities in 1980 would

get you 4.03 units in 2020. So, the average inhabitant of the planet saw a 303 per cent increase in personal resource abundance.

For example, let's look at the price of something we all enjoy; an average breakfast. Over a hundred years, the "Time Price" of a typical breakfast fell by nearly 90%. That is, for the same amount of time spent working in 1919, by 2019, you could get almost ten average breakfasts for the same amount of work. A considerable increase in abundance. That means we have more time to think, dream, innovate, create and invent. Superabundance has "created" more time.

Think about the price of flat-screen TVs. When they first arrived on the market, they cost about two- or three-months wages. Now, you can buy one for less than a week's average earnings. An abundance of TV's! According to Tuppy, when measured with the "Time Price" formula, *practically all* resources and commodities, including food, have become increasingly abundant over the last 150 years. That is a fantastic thought when considering the continuing dire predictions of catastrophe for the planet over the previous fifty years. Yet, in 1970, Paul Ehrlich claimed, "Population will inevitably and completely outstrip whatever small increases in food supplies we make. The death rate will increase until at least 100–200 million people per year will be starving to death during the next ten years." Is that rustling in the bushes a rabbit or a lion?

Even in "bad" years, abundance increases faster than the population. Here is just one of many examples from Australia – wheat. Traditionally, seed wheat produced a tall plant susceptible to rain and strong winds a hundred years ago. It was common for crops to be flattened in a storm. The wheat was also vulnerable to diseases like plant rust and often did not produce a high yield. Oh! How much has changed? Modifications to the biology of the plants now produce a shorter, more resilient, more drought and pest-resistant variety with much heavier yields. Modern technology has driven changes in equipment, such as the combine harvester, which was invented and made in Australia by the industrialist Ralph Mackay. Sowing is controlled by incredibly accurate GPS systems, increasing the plants' yield and health. The amount of human labour involved in producing a wheat crop has reduced remarkably. In short, over the last fifty years, we have grown far more grain on far less land. That frees up hundreds of thousands of hectares for a return to nature and enhances the greening of the planet we mentioned earlier.

Then, consider the vast improvements to virtually all food crops worldwide that have a similar nature and impact. And image how these changes have added to the wealth and productivity of the poorer nations, dragging them from poverty to prosperity. Marion Tuppy says, "Today, extreme poverty is no longer a global problem but is still an African problem, yet even the world's poorest region saw extreme poverty decline from 55 per cent in 1990 to 42 per cent in 2015. That reduction may seem underwhelming until we realise that the population of sub-Saharan Africa doubled from 512 million to over 1 billion over the same period.

Abundance is increasing at a faster rate than the population is growing.

Ehrlich argued that as the population increased, resources would become less abundant. Simon claimed the opposite; that a growing population would increase the abundance of resources because with every hungry mouth comes a brain capable of reason and creativity. He was right.

It is simply not true that mankind has a negative effect of this wonderful planet. Take Australia for example. When Europeans came here in the 18th century, the whole continent was the equivalent of the world's greatest managed estate! As I explained in <u>WALKING AMONG THE STARS</u>, the Aboriginal people were wonderful managers of the land. For thousands of years, they used local knowledge and ingenuity to overcome "the thorns and thistles." They used ingenious management techniques such as controlled fire to bend and transform the environment to their own benefit as well as being beneficial to the flora and fauna that provided their food and sustenance. Skills and knowledge that are sorely missed on the continent today.

It is our contention that resources grow more abundant, not despite population growth but (in large part) because of it. History shows that, unlike other animals, humans are intelligent beings capable of innovating their way out of shortages through greater efficiency, increased supply, and the development of substitutes. Over time, we have developed sophisticated ways of dealing with adversity, enhancing our chances of survival and improving our well-being. This planet rewards risk and resilience – that's how it was designed.

✳ ✳ ✳

Summary.

> *The natural state of mankind is to favour the negative over the positive. But that is not necessarily the best position. Good news is always better than bad news.*

Despite the loud and continuing predictions of doom, the world has exhibited super-abundance in almost every basic commodity calculated by the "Time Price" method. We are creating wealth, not poverty.

God charged us to flourish and multiply. We are doing both. Abundance is increasing at a faster rate than the population is growing.

CHAPTER 24

The Resilient Planet

Genesis 9:3 says:

> The fear and dread of you will be upon every wild animal, every bird in the air, every creature populating the ground, and all the fish in the sea; they have been handed over to you.

The reality is that animals, birds, fish and insects always die. The 'duality of reality' is that of the billions of species that have existed since creation; only some nine million are still alive and well. Consider this: even today, most of those species have not yet been named, which was the task given to Adam millennia ago. In a recent article, says:

> More than 99 per cent of all organisms that have ever lived on Earth are extinct. As new species evolve to fit ever-changing ecological niches, older species fade away. But the rate of extinction is far from constant. At least a handful of times in the last 500 million years, 75 to more than 90 per cent of all species on Earth have disappeared in a geological blink of an eye in catastrophes we call mass extinctions.

Undoubtedly, the number of species extinctions has been significant over the last few hundred years since industrialisation began. But the number may not be as high as the catastrophists claim, and many species, once considered extinct, are reappearing and making a remarkable comeback. For example, the Bengal Tiger, once considered on the brink of extinction, has reappeared spectacularly. Once reported to be a total population of only 2000-3000, there are now nearly 3000 in India alone. Similarly, The International Union for Conservation of Nature (IUCN) reported that the threat to Fin Whales, the second-largest mammal on the planet, and the Mountain Gorilla, that iconic symbol of human predation, had subsided somewhat. Both species were downgraded to a lower-risk category. The same is true for the Snow Leopard of central Asia, which was recently downgraded to *Vulnerable*, and the African Elephant. There are also the so-called "Lazarus Species," those thought to be extinct that have been re-discovered and thriving. In 2010, the Australian scientists Diana Fisher and Simon Blomberg reported that more than one-third "of mammal species that have been classified as extinct or possibly extinct, or flagged as missing, have been rediscovered." Between 1981 and 2001, scientists rediscovered 89 different plant species once believed to have been extinct in Australia alone.[11] The Fernandina giant tortoise, long thought extinct, re-appeared on the Galapagos Islands last year.

As Joakim Brook commented recently, "Slowly, humans are making amends for our past excesses. What's clear is that nature's resilience is stronger than we thought."

Besides, in the last 500 years, only some 80 mammals have been recorded as having gone extinct. In his book *More From Less*, Andrew McAfee, a board member of HumanProgress.org, discusses how relatively rare recorded extinctions are – with some 530 across

---

[11] Human Progress. The Return of the Dead Countering Extinction

all species in the last five centuries. More importantly, he notes, the extinction rate "appear[s] to have slowed down in recent decades; for example, no marine creatures have been recorded as extinct in the last fifty years."[12]

Many species can adapt to changed circumstances, migrate to different habitats, and, more importantly, live and operate where scientists and wildlife biologists don't directly observe them.

This is all good news, rarely reported in the Media, yet vital in achieving a balanced view and understanding of the issues. It is now apparent that humans are making amends for past excesses. Conservationists are making a concerted effort to breed known endangered species in captivity before releasing them into the wild. In New Zealand, they have declared a "Predator Free 2050" goal, which makes far more sense than an "emissions-free" goal. On the South Georgia Islands, among many, they are exterminating invasive rats and mice to benefit the indigenous species.

We are living on a most resilient planet!

You may have noticed a theme here. Many of the endangered and extinct species are endemic to islands. Matthew White Ridley, 5th Viscount Ridley, is a British author, journalist and businessman. He is known for his writings on science, the environment, and economics. He has been a regular contributor to The Times newspaper. His books have sold over a million copies. He says that, particularly in the 19th and early 20th centuries, extinctions were high because of human activity. No denying that. But he points out that most occurred on island habitats, like the Poles, Hawaii and Australia (regarded as an island by biologists) and were mainly due

---

[12] Human Progress. The Return of the Dead Countering Extinction

to the introduction of invasive species like cats, rats and other mammals like foxes.

Nevertheless, the number of species of mammals and birds that have gone extinct due to human activity in non-island habitats - is nine![13] That's right, nine! That's amazing.

> In a recent address, he says, "I want to end with a couple of anecdotes about the island of Spitsbergen near the North Pole, where I spent a summer 30 years ago with some friends. We camped next to a little Hut called Pluto Hut in a beautiful valley in Rindalen. We slept in our tents and had no guards because we didn't need to worry about Polar Bears. There were no Polar Bears on the west coast of Spitsbergen 30 years ago. One would never dream of worrying about it. Then, last summer, I inquired about renting that Hut; I don't like sleeping in tents anymore, but I thought a Hut might be all right for a week with some friends. "Let's go back there." But, no, I was told, "No, you can't use that Hut anymore; it's been destroyed by the bears!"
>
> Polar Bears are now widespread on the west coast of Spitsbergen. It has nothing to do with the fact that the ice has gone. Spitsbergen has never had ice in the summer anyway, not for 500 years. It's just an increase in numbers. The Goose colonies on the offshore islands have failed to produce a single gosling for the last five or six years because Polar Bears eat all the chicks. So, the Polar Bear numbers are right up, and so are the Ringed Seal numbers in the fiords and the Walrus numbers. The White Whale numbers have also trebled in that area.

---

[13] Matt Ridley: On How Fossil Fuels are Greening the Planet

It's not greener because most of these things are brown, but wildlife still thrives in the Arctic.

It's the same in the Antarctic. Numbers of Fur Seals and King Penguins around Antarctica have shot up. Why is that? It's elementary - we've stopped killing them. All these creatures used to be hunted for their fur, blubber, or something like that, so a massive industry was exploiting them. It's petered out now, partly because of pressure from conservationists but also because of the products it produces because we get them from fossil fuels instead. The discovery of petroleum meant that we didn't have to continue hunting whales for whale oil to the same extent. Likewise, we've all got central heating because of fossil fuels, so we don't need Polar Bear furs.

My point is that when you think about the recovery of wildlife in these areas, much of the credit must go to technology, growth and economic development. It is part of the solution, not the problem."

Matt Ridley: On How Fossil Fuels are Greening the Planet

**

Summary.

> *We humans were given dominion over the whole animal kingdom. We are as accountable for our failures as much as we are due credit for our successes. And yes, there have been plenty of both.*
>
> *The excesses of the past, as civilisation and industrialisation spanned the globe, are largely gone, as we realised the damage*

we were doing. Indeed, humanity is making amends for past mistakes.

There is no mass extinction – the whole theory of a domino-like catastrophe is a myth.

Here is the good news. Because of fossil-fuelled advances and enhancements, the greening of the planet is leading to a decreasing human footprint, resulting in a revival of some species and the flourishing of many more. Even among strife and war, there is much to commend human progress.

As well as being a source of thorns and thistles, this resilient planet is a source of joy and wonder. But it was given to humanity as a blessing. We should be worshipping the source of our blessing, not the planet. Historically, worshipping the creation, not the Creator, has never been a good idea.

CHAPTER 25

The Golden Calf.

Remember the story about Ithamar, the lonely shepherd boy who vanquished the bear, saved his flock, and became the scribe to Moses, who wrote the five books of Torah.

Three thousand men were executed that day for leading the worship of the golden calf, made from gold and other wealth that the Hebrews had bought out of Egypt. After six hundred years of slavery and suffering, Moses, led by God and with many miracles, released the children of Israel and bought them out of Egypt, promising a brighter, more prosperous future. A promise of abundance and freedom. Yet, as soon as his back was turned, they rebelled while he was shrouded by the mist that covered the mountain.

All the while, amidst all the joy and blessings, the Israelites turned to other gods in his absence. They had not bought any idols or symbols out of Egypt to remind them of their captivity. Still, they yearned for the more familiar gods of their captors. So, at the earliest opportunity, they collected the gold they had been given and made their own god, an idol, a golden calf. While Moses went to the mountain top and worshipped the living God, below, the people, created in the image of God, were worshipping an inanimate, lifeless idol.

In the midst of all our prosperity, something sinister is happening today, a deception promising a similarly disastrous result. It again exposes the reality of duality – a multitude who follow a materialistic view of science, that the whole universe is explained merely by the natural arrangement of inert, uncomprehending atoms, that nothing exists outside or beyond that. Yet, they refuse to see another dimension; that all of creation, everything physical, every atom, has at its very source a basic substrate – the mind. A deception that refuses to acknowledge an all-encompassing mind, a superior intelligence, creativity, and design. To recognise the Creator manifest in the natural world, the same God that blessed Adam and Eve and sent them out to overcome the thorns and thistles, to endure the strife and struggle, to have dominion, to relish the freedom, to create, to innovate, build, and enjoy this wonderful planet.

And yet, the duality of reality, the conflict, is the growing movement to grant spiritual status to that so-called natural world. To worship creation, not the Creator, to call on the planet or the Universe, rather than the magnitude of the Maker. Some are calling it a new religion, a false religion of the masses. There's nothing new about it. It is the oldest religion, dating far back before recorded time. The pagan worship of the planet, of creation, of Mother Earth, and ultimately of Gaia, the primordial goddess and Earth mother. It is said that Gaia emerged from Chaos and is considered one of the first beings to exist. It is first and foremost a pagan religion, stalking the earth today rampant, foreboding – and popular. It has many names and manifestations but has wormed itself, often nameless and unidentified, into modern society like some primeval nightmare.

For millennia, mankind struggled with the thistles and thorns. Universal poverty measured human progress, barely raising the abundance graph from a flat line. Then suddenly, like Hebrews being delivered from Egypt, humanity exploded into a life of abundance that could scarcely have been imagined a couple of hundred years earlier. We have been truly blessed. The whole planet *is being* blessed. True, there is still strife and war – but far less than previously, yet humanity is continually and bountifully blessed. And our response? To worship an unfeeling, inanimate, lifeless idol, an earth goddess!

The WOKE environmental movement is a counterfeit, faithfully copying the tenets of Judeo Christianity. They believe that before mankind, nature was once perfect and should be returned to that state, thus reflecting the Garden of Eden before the fall from grace. They believe we live in a corrupt world surrounded by pollution and environmental damage caused by humanity. They believe there is a day of judgment coming for us all. We are sustainability sinners doomed to destruction unless we seek salvation in the environmentally perfect state of sustainability. This is the religion of Mother Nature, of Gaia.

We have created a complex and abundant civilisation in the past few hundred years. We are above the animal kingdom, endowed with the *neshama*. This purely spiritual characteristic leads to a culture of sharing knowledge that, in turn, invites invention and innovation. That springs from a free society and creates a gentler, more prosperous society. Marion Tuppy says, "Population growth and freedom are crucial to that positive feedback loop. It is free people, not machines or deities, who generate new ideas, and it is free people who test those new ideas against other people's ideas in the marketplace. The process of knowledge and value creation is at the heart of humanity's moral and material progress."[14]

---

14 Superabundance. Marion Tuppy.

You will find a spiritual core at the heart of the environmental WOKE movement. Yet deep down, it displays all the attributes of Judeo-Christian morals and ethics embedded in Western Culture. It is a deception. Marion Tuppy continues … "In the Judeo-Christian tradition, human problems stem from our failure to adjust ourselves to God, while in the apocalyptic environmental tradition, human problems stem from our failure to adjust ourselves to nature. In this secular religion, God is replaced by nature, and the priesthood is replaced by scientists who are tasked with interpreting the natural order of things."

The gospels of this pagan religion are Climate, Gender, Diversity, Equity and Inclusion. We have already covered the response to a changing climate. So consider this: "God created humankind in His own image – male and female, he made them."[15] He made humankind in all shapes and sizes, in different colours, with differing characters and life experiences. As His image bearers, he created diversity! There is nothing so diverse as humankind. Equity requires that all are equal in all respects – wealth and privilege. Quoting the Torah, Jesus said, "The poor will always be with you." Deuteronomy 15:11 says:

> There will always be poor people in the land. Therefore, I command you to be openhanded toward your fellow Israelites who are poor and needy in your land.

Jesus also taught us to treat others as we wish to be treated. That is genuinely inclusion.

**

Ch15 Genesis 1:27

When Paul Ehrlich published his book, The Population Bomb, in 1970, most commentators agreed with his catastrophic predictions that we would run out of food and resources within the decade because of population growth. Instead, despite the world's population doubling since then, food production has stayed ahead of population growth, famine is less common, and resources are even more affordable. Despite all evidence to the contrary, many, particularly young people, are drawn to the predictions of doom like moths to a candle.

In fact, the opposite is true – it is now widely recognised in the scientific community that the world's population is in decline. That decline will reach catastrophic proportions in fifty years' time when there are not enough of working age to support the older population. While the world is still discussing climate change and population growth, this is now widely acknowledged as a significant issue. But why is it happening? Statisticians and economists tell us that the proportion of women having multiple children hasn't changed much. Neither has women having just one child dropped significantly. The real change driving this phenomenon globally is the number of women not having children; this has risen exponentially.

The "replacement rate" is the number of children per woman required to stabilise the population. That is, there are 2.1 children on average per woman. This has dropped in developed Western countries for about 50 years; for instance, in the USA, (1.66) children per woman. In Australia, (1.63) Japan, (.78) China, (1.18.) Only Africa, which is still mainly struggling with energy and poverty, is showing any population growth. (4.47)

We have seen that population growth and freedom, hand in hand, promote innovation and invention, which in turn leads to human progress. Food and resources lead to population growth in a free society, creating abundance. Conversely, de-population does not. Traditionally, it has led to war, famine and poverty. "God said to them, "Be fruitful, multiply, fill the earth and subdue it."[16] He has never rescinded or corrected that command.

One of the saddest things I see in the world today is young women refusing to have children, "to save the planet!" Driven by the totally negative news cycle, by catastrophism and crisis, many young women are afraid to bring children into the world. This is a disaster in a world crying out for babies.

> Equally appalling is the number of abortions that are being performed, currently 75 million globally every single year. Let's say half of those unborn would have been women. Imagine the number of babies they could have produced in a year, in a decade? As a Christian, I believe abortion should be safe, rare and legal. Let's oppose abortion except on the grounds of medical emergencies. Psalms 139:13-16 says:
>
> > For you created my inmost being;
> >
> > You knit me together in my mother's womb.
> >
> > I praise you because I am fearfully and wonderfully made;
> >
> > Your works are wonderful,
> >
> > I know that full well.
> >
> > My frame was not hidden from you.
> >
> > when I was made in the secret place,

---

[16] Genesis 1:27

when I was woven together in the depths of the earth.
Your eyes saw my unformed body;

Summary.

Like the children of Israel, once free of slavery and being led toward a life of freedom and abundance, many are now turning to the pagan gods of old, like Gaia, the Earth mother. This has never been a good idea.

Neither has worshipping the WOKE gods of Climate, Gender, and the socialist ideas of Diversity, Equity and Inclusion ever led to anything good and prosperous.

The real crisis today is one of de-population, a phenomenon that is growing globally.

CHAPTER 26

The Third Day

One evening in August 2024, I was suffering from a severe staphylococcus infection that attacked my heart. I had a series of cardiac arrests, differing from a heart attack, in that it's more of an electrical communication problem. I had my first cardiac arrest in the ambulance parked outside my front gate. That was supposed to be it – the chance of surviving another is remote. I had a second and stayed alive until I finally got to the Emergency Room at the hospital. I had another – I recovered to see a group of doctors and nurses surrounding my bed, yelling and screaming at me. This wasn't supposed to be happening. I felt OK and claimed I had just fallen asleep. They just pointed to the flat line on the monitor. They were amazed I had recovered.

The doctors realised I could tell when an attack was about to happen and encouraged me to raise my hand as a signal. I lost count of how many times I raised my hand that night and recovered. I knew that every time could be my last. My family were in a waiting room. When I talked to them later, they said a Code Blue alert had repeatedly gone on all night. I died many times and came back to life. Finally, they changed my antibiotic in the early morning, and it worked. I have had no cardiac arrests since! To finish things off, I had a CT scan, and before I left on the Air Ambulance to Melbourne, I was told I also had lung cancer! My family were advised I probably wouldn't survive the trip – but I did, and here I am today, eight months later and flourishing. Thanks to a fantastic medical team.

To say that my life has changed would be an understatement. I feel like a totally new man, which, in a sense, I am. I continually thank God for my new life and the opportunity to finish this book. People often ask, "What was it like to be dead?" I tell them it was exactly as I expected, just like falling asleep, just as the Bible says. Except, I kept waking up.

> For this we tell you by the word of the Lord, that we who are alive, who are left to the coming of the Lord, will in no way precede those who have fallen asleep. For the Lord himself will come down from heaven, with a loud command, with the voice of the archangel and with the trumpet call of God, and the dead in Christ will rise first.
>
> 1 Thessalonians 4: 15-17

Do I ponder why this happened to me? Yes, constantly. Yet, I am overjoyed by the happiness my recovery has meant to my friends and family. Perhaps it has allowed me to write this final story with empathy and a small grain of understanding?

***

The Man was battered, bruised and broken as they roughly laid him down on the wooden cross and nailed his wrists and ankles to the timber. They fixed a sign saying "King of the Jews." He did not yell or complain when they raised him up for all to see, yet his agony must have been unbearable. His followers looked on in horror at his bloody pain and anguish. Finally, He said, "It is accomplished," and He gave up his Spirit and died. The sun went dark. They took his mutilated body away and buried Him in a tomb.

Many rejoiced, including the Principalities and Powers,[17] *the Spiritual Rulers of this world. For three days, for 72 long hours, they rejoiced and celebrated their victory over the Son of Man.*

But then, without warning, the party was over!

On the third day, He awoke, He came back to life, and He and rose again, triumphant and victorious!

***

For God so loved the world that he gave his only and unique Son, so that everyone who trusts in him may have eternal life, instead of being utterly destroyed.

John 3:16 Complete Jewish Bible.

Amen.

---

[17] For our struggle is not against flesh and blood, but against the rulers, against the authorities, against the powers of this dark world and against the spiritual forces of evil in the heavenly realms. Ephesians 6:12 NIV

CHAPTER 27

In Conclusion.

In my introduction, I quoted the words of the great British philosopher Sir Roger Scruton, who wrote:

> *"Anybody who goes through life with an open mind and heart will encounter moments saturated with meaning but whose meaning cannot be put into words. These moments are precious to us. When they occur, it is as though, on the winding, ill-lit stairway of our life, we suddenly come across a window through which we catch sight of another and brighter world to which we belong but cannot enter. There are many who dismiss this world as unscientific fiction. I am not alone in thinking it real and important."*

I invited you to join me on a journey – *"in search of another and brighter world, to which we belong but cannot enter."* Some say the journey is more important than the destination – I agree and trust that you have found it so and that we will continue on this voyage of discovery. Exploring ideas, considering and debating new concepts and ways of thinking.

The time I spent in ER, suspended for a night between life and death, *"were moments saturated with meaning but whose meaning cannot be put into words."* When I first wrote that nearly two years ago, I had no idea of the implication those words would eventually hold. I have experienced a significant truth – that we cannot see tomorrow, we cannot see today, or even look forward confidently to the next minute, even the next second. We cannot see that *"to which we belong but cannot enter."*

And yet, we are, above all, *"spiritual beings having a physical experience."* In the Spirit, we can sense and experience that other world, looking forward to the future, as though peering through a darkened glass. The Bible, the inspired Word of God, is our way into that other world, the Holy Spirit our guide and companion along the highways and byways of our journey together.

Is there meaning to our lives that transcends the splendour we see about us? Something that goes beyond the physical? And if so, how can we probe that meaning? So much of life seems to hang upon uncontrollable events.

***

The Bible contains amazing literature, beautiful poetry, and memorable stories. It is also historically factual, reliable and verifiable. Science and the Bible are complementary works, with the Bible ascendant. Nature should be read as a book, revealing God's word and plans for creation - the duality of reality. Modern science was born in the crucible of Western civilisation. Contrary to popular belief, most early scientists believed in the Bible.

The Bible has primacy over everything. But it is scripture, *not* our own interpretation, that provides revelation. The bible says that there was a beginning in the beginning! It is silent about how or when, except it came about through God's word. The scientific theories about how and when that happened are constantly changing, as are estimates of time and duration. When the truth is arrived at, if ever, science and the Bible will agree.

The Luckiest Planet

The Bible is silent about the length of time between Creation and Adam, except that it differentiates between the time leading up to Adam being regarded differently. However, we have established in a methodical scientific way how six 24-hour Earth days can also be viewed as billions of years by the stretching of time caused by the universe's expansion. The duality of reality. This satisfies both our scientific understanding of the natural world and the theology of the Bible. The opening passage of the Bible makes the most astonishing claim in the history of humanity. It describes the first and foremost of many miracles, as God enters his Creation- with intent. The universe was designed for life from the moment of Creation. As was Planet Earth, the luckiest planet in the universe. The odds against life forming and flourishing on this tiny planet are so high as to be beyond calculation!

※ ※ ※

According to the fossil record, life appeared as basic single-cell bacteria soon after water appeared in liquid form on planet Earth. Primitive but incredibly complex. We do not know how that happened, but we may one day. But we do know why that happened. God spoke to the planet, our Earth, and told it to "bring forth life," using all the elements necessary for life to flourish. He did not specify a timeline, for God does not need one.

For millions of years, nothing much really happened.

Then photosynthesis from the living organisms eventually purified the previously dense atmosphere, and the sun, moon, and stars became visible from Planet Earth. But then, billions of years after the Big Bang, an amazing event occurred that would change life on Earth forever! Soon after the arrival of water on the planet, life started abruptly, something science has failed to explain. Life appears suddenly, mysteriously in the fossil record. Simple single-celled life yet incredibly complex, based on "Information."

There was no slow evolution – it just started, with complex cells and organisms soon afterwards. Then, 530 million years ago, it exploded. But, since then, no new phyla (body types) have emerged. The only changes have been made through natural selection.

***

The Earth brought forth life, plants, creatures, and animals. One of those species was hominids, homo sapiens. God chose Adam and Eve from them and created a new Spirit in them, the *neshama*, a New Creation. Spiritual beings created in the image of God. He blessed them and separated them from all other living things. He gave them dominion over the whole planet and every living thing.

And He saw it was very good.

There is no evidence in the fossil record to suggest a link between the ape family and homo sapiens. About 6000 years ago, Adam and Eve were chosen by God to become the New Creation. They were the first human beings. They were set apart, above the animal kingdom, infused with the *neshama*. They became communicating spirits. They had a personal relationship with God.

They were destined for eternal life.

So are you!

***

In the last century, Darwin's theory of evolution reigned supreme in scientific and religious circles. Recent advances in science and biology refute the bedrock of the theory that natural selection leads to new species. There is no proof of that in the fossil or natural world. The old theories and assumptions of evolution that were not based on empirical evidence are being swept away by a new wave of scientific knowledge and understanding. There is a growing awareness that Darwin's theory of progressive evolution is skating on thin ice. More and more scientists realise that Intelligent Design is necessary to better understand our natural development. As we look at the natural world through the perspective of both the Bible and Science, it becomes increasingly apparent that a mind is at work. An intelligent being. A designer. Just a simple reading of the Bible will identify that Mind, that Intelligent Designer. That Creator.

The Torah and the Genesis chapter were written 3500 years ago. During that time, the original Hebrew language changed many times. Remember, when written, it was for an ancient, mainly nomadic illiterate people. Things are not always so clear in today's language. So yes, dinosaurs were mentioned in the Bible. But to discover that takes some digging.

※※※

In Part 2, we spoke about the three-legged stool of human misery: Slavery, War and Discrimination.

Slavery has always been a part of the human condition, sometimes an essential and irreplaceable part of a nation's economy.

It was only through the growing influence and understanding of the Bible that people came to realise that slavery was evil and not part of God's plan for humanity.

That's when things started to change.

War and violence have also been part of the human character since the beginning – but that does not make it just or defendable.

There is much spoken about racism today. Much of it is the loud beating of a very hollow drum. If you want to stop racism, stop being racist.

These have been the days of my life so far. I have witnessed health and happiness, death and destruction, tragedy and terror in those days. Much of that has been imposed by humans upon humans. This can be changed, one person at a time, one day at a time. It can start with you and I.

***

It is God's blessing that we be fruitful and prosper; human flourishing is God's idea! That promise and blessing have never been revoked. It is current today.

Our responsibility <u>as a race</u> is to have children and for them to have children. God's will has us populate the whole earth, and God has <u>created the world significantly to sustain us.</u> But it will take effort and sweat to defeat the thorns and thistles. It won't be easy.

Our obligation to the planet? To manage the whole planet so that humans flourish and prosper sustainably. To ensure that <u>all life</u> is properly cared for and protected. God's plan is for humans to be in charge of nature and work with nature – not the other way around!

We are collectively responsible for humanity –all made in His Image. In other words, how we govern and manage our people and treat other nations must be done with love and compassion. In God's eyes, all humans are created equal and should be treated the same way.

The Luckiest Planet

***

There is a lot of talk about climate catastrophe today. It is mostly talk, with no proper understanding of the climate or what a real catastrophe actually looks like. A real climate catastrophe could happen at any minute, without warning, and turn our planet into darkness and desolation.

Let's be thankful for the wonderful, stable climate we enjoy today on this luckiest of all planets!

Carbon Dioxide is not the life-threatening substance we have been led to believe. Life as we know it ceases to exist at 150ppm (parts per million) of CO2. Before the industrial revolution, at 240ppm, we came perilously close to that. Thanks to human impact, we now have a much more favourable climate for plant life. And warmer temperatures are improving, not degrading the lives of millions of people in the colder climates. Historically, far more people have died from cold than from heat.

We can cope with these changes incrementally with the advances in human knowledge and innovation over that same period. Climate-related deaths have decreased remarkably.

The climate is comprised of an amazingly complex set of variables. It is not possible to model future predictions accurately. Perhaps in the new age of AI, that will change. But not yet.

We are still in an Ice Age. The rise in temperature (1°) and sea level rises (2cm per annum) over the past 150 years are welcome and well within the anticipated normal levels of climate variation. In the 4.5-billion-year history of this luckiest of planets, humans have seen far more impressive and catastrophic events than we are experiencing now.

The solution to climate change is not poverty – it is to adapt to the changing climate just as we have been doing for thousands of years.

Fossil-fuelled machines and technology have reduced climate-related deaths by 98% and will continue to reduce our "climate danger." Some 3 billion people are without life-changing electricity. Billions are being denied the life-empowering, fossil-fuelled electricity that we enjoy because of an "addiction" to unnecessary expensive climate mitigation ideas and policies. Together, we can change this. Put simply, cheap, affordable energy drives wealth and prosperity. And the opposite is true.

***

The natural state of mankind is to favour the negative over the positive. But that is not necessarily the best position. Good news is always better than bad news. Despite the loud and continuing predictions of doom, the world has exhibited super-abundance in almost every basic commodity calculated by the "Time Price" method. We are creating wealth, not poverty. God charged us to flourish and multiply. We are doing both. Abundance is increasing at a faster rate than the population is growing.

We humans were given dominion over the whole animal kingdom. We are as accountable for our failures as much as we are due credit for our successes. And yes, there have been plenty of both.

The excesses of the past, as civilisation and industrialisation spanned the globe, are largely gone, as we realised the damage we were doing. Indeed, humanity is making amends for past mistakes.

There is no mass extinction – the whole theory of a domino-like catastrophe is a myth.

Here is the good news. Because of fossil-fuelled advances and enhancements, the greening of the planet is leading to a decreasing human footprint, resulting in a revival of some species and the flourishing of many more. Even amongst strife and war, there is much to commend human progress.

***

As well as being a source of thorns and thistles, this resilient planet is a source of joy and wonder. But it was given to humanity as a blessing. We should be worshipping the source of our blessing, not the planet. Historically, worshipping the creation, not the Creator, has never been a good idea. The question remains – who do you serve, the creation or the Creator?

THE END

www.ingramcontent.com/pod-product-compliance
Lightning Source LLC
Chambersburg PA
CBHW071957070526
44583CB00015B/1225